Wells and Septic Systems

2nd Edition

Max and Charlotte Alth

Revised by S. Blackwell Duncan

SECOND EDITION
SECOND PRINTING

Library of Congress Cataloging-in-Publication Data

Alth, Max, 1927–
Wells and septic systems / by Max and Charlotte Alth ; revised by S. Blackwell Duncan.—2nd ed.
p. cm.
Includes index.
ISBN 0-8306-2137-7 ISBN 0-8306-2136-9 (pbk.)
1. Wells—Design and construction. 2. Wells—Maintenance and repair. 3. Septic tanks—Design and construction. 4. Septic tanks--Maintenance and repair. I. Alth, Charlotte. II. Duncan, S. Blackwell.
TD405.A536 1991
628.1'14—dc20 91-21405
CIP

TAB Books offers software for sale. For information and a catalog, please contact TAB Software Department, Blue Ridge Summit, PA 17294-0850.

Acquisitions Editor: Kim Tabor
Book Editor: Debra Marshall
Director of Production: Katherine G. Brown
Book Design: Jaclyn J. Boone

Dedicated to
Erin
Shannon
Kim
Simon
Michele
Darcy
Michael
Charlotte
and all other lovers
of pure drinking water

Contents

PART 2
WATER WELLS

Introduction

You would like to own your own home, either a year-round home or just a home for the summer. Not too many years ago many of us could realize this dream only after a dozen years or so of planning. We could work hard, scrimp and save, and accumulate sufficient money for a down payment. Eventually, we would own our own home free and clear.

But today, the difference between our income and the sum needed to purchase a home is several times greater than ever before. One way to reduce the sum and also the time it will take to raise a down payment is to purchase property that lies beyond the reach of municipal water and sewer lines. Rural and semirural property is generally less expensive than city or suburban property, and the taxes lower.

Most of us have lived in homes that have a free supply of water at the end of a faucet and a toilet bowl that flushes with the turn of a handle. It can be disquieting to think of depending upon the vagaries of one's own water and sewage systems.

But there is no need to feel uneasy. Millions of American homes have their own water and sewage systems. The owner-builders of millions more install their own systems to reduce the cost of building their homes. The systems work and are dependable.

This book describes and explains how these systems work, how to do much, if not all, of the work yourself, and how you can maintain the systems with a minimum of time and effort. Browse through the book to get a good idea of what water wells and septic systems are all about, then go back and digest the portions that interest you most. But by all means, before you start construction, or even purchase a building site, carefully consider the details of percolation tests, spelled out in chapter 2, and also the discussion of water rights and water contamination in chapter 12. An understanding of these subjects and their importance could save you a lot of grief later.

Part I

Septic systems

Chapter **1**

How septic tanks work

Most populated areas are served by municipal sewage disposal systems, and all a home-builder has to do is tap into the pipeline. But in rural and semirural areas municipal sewage service is not available, and home owners must install what is called an on-lot subsurface sewage disposal system, an individual household waste treatment system, a septic tank sanitation system, or simply—a septic system.

Septic systems consist of an underground tank and a leach or drain field that work to cleanse and purify household waste water. This practical, safe, effective system is currently used by more than half of the rural and semirural homes in the United States and Canada, and perhaps 15 percent of urban and exurban American homes. Outside of a biennial cleaning and some care, the system usually requires no attention (there are exceptions caused by unusual site conditions). The tank can last indefinitely, depending upon its construction. The leach field will probably have to be treated or replaced after 15 to 20 years of service, occasionally sooner. The system is invisible, odorless if installed and functioning properly, and it rarely has moving parts. Figure 1-1 is an outline of a typical installation.

Sewage flows by gravity from the house to the tank. (Occasionally sewage must be mechanically pumped upward into a tank at a higher level.) There the sewage stands for the time needed for anaerobic bacteria to break down the solids. Incoming sewage displaces a like quantity of liquid, which flows from the tank outlet by gravity (again, it can be pumped if necessary); this displaced fluid is called the *effluent.* The effluent enters buried leach field pipes where it seeps into the surrounding soil. This action filters the liquid, while aerobic bacteria further break it down into various nutrients and chemicals that support plant life. Some of the moisture may transpire into the air above, de-

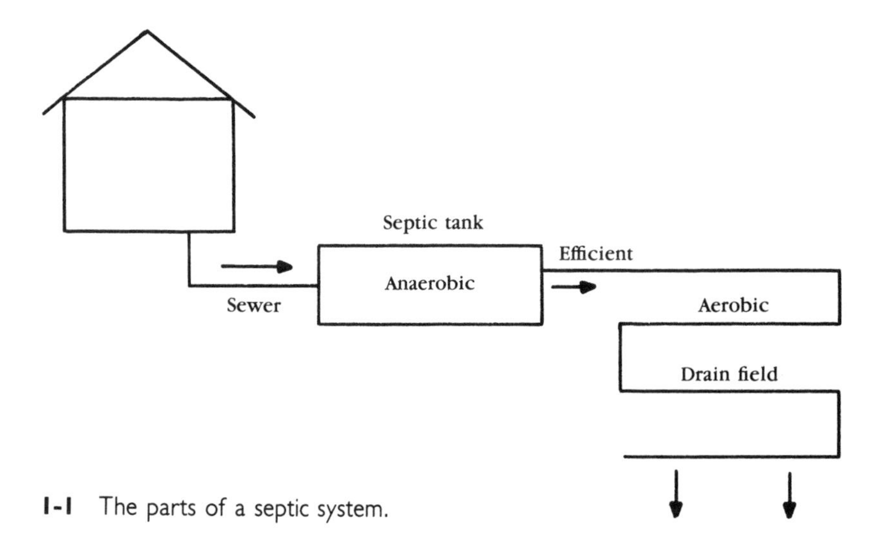

1-1 The parts of a septic system.

pending upon leach field design, and the remaining effluent eventually reaches an aquifer. By this time it is water, fit for human consumption.

BASIC DESIGN

The average individual uses anywhere from 50 to 350 gallons of water per day for drinking, cooking, washing, flushing the toilet, and other household purposes. Except for homes that have provisions for separating water containing no solid human wastes from that which does, 99.9 percent of sewage is water. The remaining 0.1 percent is solid. Of this, roughly 80 percent is organic and the balance is inorganic. The organic substances derive from feces, detergents, soaps, urine, and food bits produced by garbage grinders, or which have entered the septic in other ways. Water softeners, borax, paint, photographic chemicals, household cleaners, and the like are the sources of the inorganic solids found in household sewer lines.

Roughly 50 percent of the water coming down the sewer pipe is only slightly polluted. This is water from the kitchen sink, the shower and the bathtub, and is called *gray water*. The balance is water used to flush the toilet. Because it is heavily polluted, it is called *black water*.

In terms of quantity, there are few truly dangerous substances in black water. Humans excrete between 16 and 100 ounces of feces every day on average. This is carried through the house main drain by 30 to 300 gallons of water per person per day. Thus, the system handles a large quantity of liquid containing a small percentage of solid.

The tank and its function

The tank is a large watertight, light tight and relatively air tight container positioned beneath grade level. Sewage is carried to the tank by an extension of the house main drain, the sewer line. A second pipe leads from the opposite side or end of the tank to the leach field. The field consists of a number of loosely joined pieces of clay pipe (drain tile) set in a bed of gravel, which is in turn covered by a layer of earth. More commonly nowadays, 10-foot lengths of perforated plastic pipe are laid end to end in a gravel bed. The soil above provides a bed for grass that is indistinguishable from any other on the property.

Water movement within the leach field is by gravity alone. There is no odor or unusual dampness whatsoever, and the area can be used for any purpose except the passage of vehicles or heavy equipment, or for grazing by livestock (because they compact the soil, cutting down evaporation, and potentially can cause structural damage to lines close to the surface).

The tank acts as a stilling pond and as a breeding ground and incubator for beneficial anaerobic bacteria, yeasts, fungi, and actinomycetes. When the comparatively rapidly moving waste liquid leaves the sewer pipe and enters the tank, movement almost ceases. There is only some slight, temporary surface rippling. This stilling serves two functions: The solids sink to the bottom, and the anaerobic creatures, which grow much better in still than in moving water, thrive.

These microscopic creatures attack and digest the organic solids as they sink to the tank bottom. In the process, methane and other gasses are produced. The gasses bubble to the surface, bringing along fine particles of solid matter which, together with oils and grease, form a scum over the surface of the liquid. This helps reduce movement of the liquid, and further insulates the anaerobic bacteria from air that seeps into the tank. The underside of the floating layer of scum makes an ideal breeding place for the anaerobic community.

As long as there is organic matter in the tank and its temperature does not drop below freezing, the bacteria continue to eat, multiply, and break down the organic material into its constituent elements. The bacteria, as well as the yeasts, fungi, and actinomycetes, work comparatively slowly—faster when warm and more slowly when cold. Roughly, they take 10 to 20 times as long to convert organic matter to soluble nutrients as do aerobic bacteria.

Anaerobic bacteria produce only about one-tenth to one-fifteenth the heat the aerobic organisms create. The anaerobic bacteria do not completely transform organic matter into its component elements. That job can only be performed by aerobic bacteria in the presence of sufficient oxygen.

When attacking an organic substance, anaerobic creatures produce water, ammonia, hydrogen sulfide, phosphates and heat. Physically, the change is from a solid to a liquid. What the anaerobic community cannot digest—pieces of stone, plastic, comparatively large, dense pieces of bone and wood—remains in the tank and must eventually be removed by pumping. With a good-sized tank and a little care, that need not be done more than every other year.

THE LEACH FIELD

The effluent that flows from the tank outlet is still *septic*: it contains substances that promote the decomposition of vegetable and animal matter as well as pathogens, bacteria, and viruses harmful to man and beast. The effluent cannot just be dumped. It must be treated, but this can be done so easily that one wonders why it took man so long to discover and apply the method. Simply stated, when the effluent discharged from the tank percolates through 4 feet of soil, in theory it becomes pure enough to drink.

In practice, nobody knowingly depends upon a mere 4 feet (a distance on which authorities differ) of filtration to purify drinking water. Generally, a minimum distance of 60 feet is required, and even this may not do the job. Effluent should not be discharged into a leach system located less than 60 feet from a well, spring or other body of water. Other considerations regarding the placement of a leach field in relation to a water supply are discussed in Chapter 2.

THE WORLD UNDER OUR FEET

Soil is soil to the naked eye, and with the exception of earthworms, moles, gophers, and the like, many people regard the soil as uninhabited. This perception is entirely false. A fertile acre can house half a ton of worms. And if you bothered to count, you would find some 5 million bacteria in a single teaspoon of ordinary soil. Viruses, the smallest creatures found in the soil (about 0.02 microns in size), enter the bodies of microbes and eat their protoplasm. While this is not good for the microbes, it is not necessarily bad for our purpose.

When times are very good for bacteria—when the temperature is high and there is plenty of air and food—they multiply rapidly and form thick gelatinous colonies in the soil. The gelatin reduces the flow of effluent through the soil, thus reducing the effectiveness of a leach field. By killing off some of the bacteria, viruses slow the growth of bacterial colonies. The bacteria are also destroyed by amoebas, which can be described as microscopic jellyfish. The amoebas are in turn preyed upon by nematodes, which eat amoebas and microbes.

Earthworms make comparatively large holes in the soil; nematodes make smaller holes. Together, worms and nematodes aerate the soil, providing the oxygen needed by aerobic organisms to do their part in the cycle of sewage purification.

The bacteria feed upon and digest the organic waste present in the effluent. In effect, the aerobic bacteria "cook" and chemically oxidize their food. Their wastes are soluble, stable chemical compounds that are harmless to man and beast, and are food to plants. Without the bacteria, fungi, and the rest of the microscopic creatures of the soil, plants would soon starve to death.

Aerobic creatures attaching the effluent from a septic tank produce the same compounds produced by anaerobic creatures in the septic tank, and heat. Although the heat produced is comparatively slight, it is sufficient to encourage the growth of plants. Nature has set up a symbiotic balance between man and the animals, the microscopic creatures in the soil and plants.

The soil and the community of creatures living in it combine to destroy pathogens (disease-causing bacteria, viruses, and the like) in six ways:

- Soil bacteria, fungi, and other microbial organisms produce antibiotics, such as penicillin, that destroy pathogens.
- Temperature, acidity, and moisture within the soil is so different than that found in human and animal bodies that the pathogens soon die.
- Protozoa in the soil prey upon bacteria and the viruses accompanying the bacteria.
- Soil organisms compete with pathogens for food, thus starving them out.
- The soil acts as a filter and prevents the larger bacteria from traveling very far.
- The soil, especially the clay, absorbs the viruses and locks them in place, preventing them from moving on.

FILTRATION

The effluent from the septic tank is carried into the leach field through perforated plastic pipe or short lengths of loosely joined clay pipe. The liquid flows out of the pipe and through the coarse gravel in which the pipe is bedded. From the gravel, most of the liquid passes into the soil by gravity, while some moves out and up through capillary action. Some of this liquid will leave the soil entirely through evaporation (FIG. 1-2) and some will leave by transpiration—a process wherein the roots of plants take up water and the water evaporates from the plants' leaves.

The soil acts as a filter, but simply passing septic water through several feet of soil will not physically filter out any but the larger particles. A purely physical filter will not remove viruses, but soil does because its action is far more complex and effective than a mere physical filter.

Soil is mainly composed of tiny pieces of stone and humus (decayed organic matter, such as leaves). When the pieces of stone are very fine the soil is called *clay*. When the particles are a little coarser, it is called *silt*. Coarser still and it is *sand*, followed by *gravel*.

Every particle of stone in the earth carries a minute electric charge that performs great and wondrous service. All pathogens have a coat of protein. In most soils, this coating somehow produces an electric charge of a polarity opposite to that of the soil particles. Because pathogens are submicroscopic viruses, they cannot resist the pull of the electric charge on the particles of stone. They are attracted to the stone and held in place. Pathogens away from their hosts die fairly quickly.

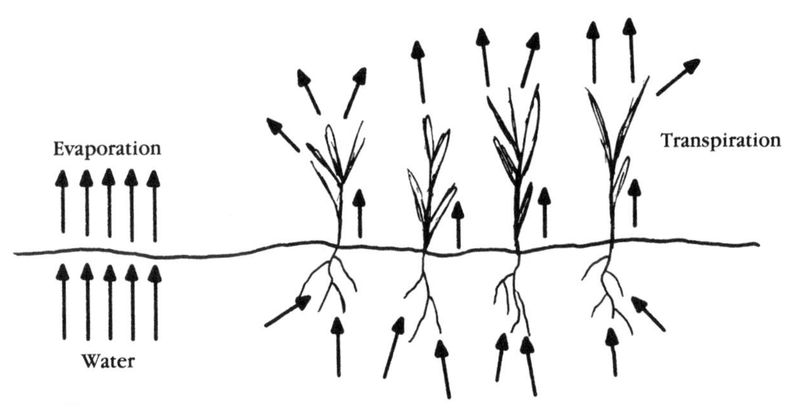

1-2 Evaporation occurs when moisture in the soil passes into the air as water vapor. Transpiration occurs when water in the soil is drawn up by plants and emitted into the air by their leaves as water vapor. Because the root system of a plant has a tremendous total length and surface area, and the leaf surfaces have great exposure to the sun and air, transpiration can easily account for the dispersion of a thousand times more water from a given soil area than the soil surface area alone can manage.

The finer the particles of stone, the greater their total surface. Clay, which has the greatest surface area, develops so much attractive power that pure clay will cleanse water of whatever viruses might be present in less than 4 inches of travel. In other words, pass polluted water through 4 inches of clay and it will be purified and fit to drink.

In addition, the electric charge of clay, and all larger particles of stone to a lesser degree, precipitates such dangerous chemicals as strontium, rendering them harmless. It also reacts with proteins and nutrients so they will not wash away and further react with lime to change soil acidity.

One point needs clarification. No one depends on just 4 inches of clay to purify the effluent from a septic tank, because clay alone cannot be used as a filtering medium. It packs so tightly it becomes nonporous. The rate of water movement through pure clay is so slow you can form a bowl from wet clay and use it to hold water. You cannot install a leach field in an area of pure clay, because water won't leave the drain lines.

The septic tank household waste disposal system provides a satisfactorily extensive purification and filtration system. During the time that household sewage remains in the septic tank, heavy particles sink to the bottom where anaerobic bacteria attack and digest the organic material. The resulting effluent, as it is displaced by fresh incoming sewage, flows into the leach field, where aerobic bacteria convert the remaining organic material to soluble, stable plant nutrients.

As the effluent flows through the earth it is filtered. Large particles are stopped by grains of clay and sand. Viruses are attracted to the surfaces of stone particles, stick, and die. Much of the now purified water finds its way downward to a water table; some returns to the surface and is dissipated through transpiration and evaporation.

During the life of the septic system, which can be 20 or 30 years or even more, there is no odor. The leach field is invisible, as is the tank, except perhaps for one or sometimes two inspection and cleanout hatches. The tank life span is unlimited if it is made of concrete or fiberglass.

The life span of the leach field depends upon its use, size, and the kind of soil that surrounds it. When the soil immediately around the pipes becomes saturated and clogged with fine particles of solid matter, the soil loses its ability to function as a filter. The effluent cannot flow easily from the lines and disperse.

As a result, the effluent will back up and appear on the ground surface, causing a distinctive and unpleasant odor. Thus, the system has a built-in alarm that warns the homeowner of trouble. Note, though, that the appearance of water and even an odor above the leach field is not a certain indication of a saturated field. Other conditions can also bring this about.

When a drain field has reached the end of its useful life, it can be restored and made an effective filter again by removing the old gravel and the clogged earth immediately adjacent, treating the nearby remaining soil, and replacing the gravel field and covering it with a layer of fresh soil.

Chapter 2

The percolation test

*T*he construction of a septic system must begin with a preliminary assessment called a *percolation test.* This test is usually conducted under the guidance of state or local health or building department personnel. The key to the system is adequate drainage and dispersal in nearby soil of the large quantities of liquid effluent. For adequate drainage and dispersal, the soil must have characteristics that enable moisture to seep more or less continuously, or *percolate*, through it reasonably freely. In other words, the soil must have an acceptable percolation rate. This rate is determined by on-site tests at the proposed drainage field location before a system can be installed. If the test results are unsatisfactory, health and zoning regulations usually forbid building a septic system on the site. Even if it can be built, the system will not function for more than a short time.

A satisfactory percolation rate is of paramount importance in your selection of a piece of property on which to build a house. So much so, in fact, that property sales are often made contractually contingent upon an acceptable percolation rate being obtained, and this is a smart way for any potential buyer to proceed. Unfortunately, many innocent buyers have been caught unawares, unable to build because a septic system could not be successfully operated on their land. Bottom line: No perc, no house.

LOCATING THE LEACH FIELD

Along with soil characteristics, a leach or leaching field (FIG. 2-1), or drain field as some people call it, is a key to the effectiveness of local sewage treatment. You can position the septic tank almost anywhere, and if the building is higher than the tank and the tank is higher than the leach field, the system will work to that point. But the leach field must drain—that is the controlling factor in the entire operation.

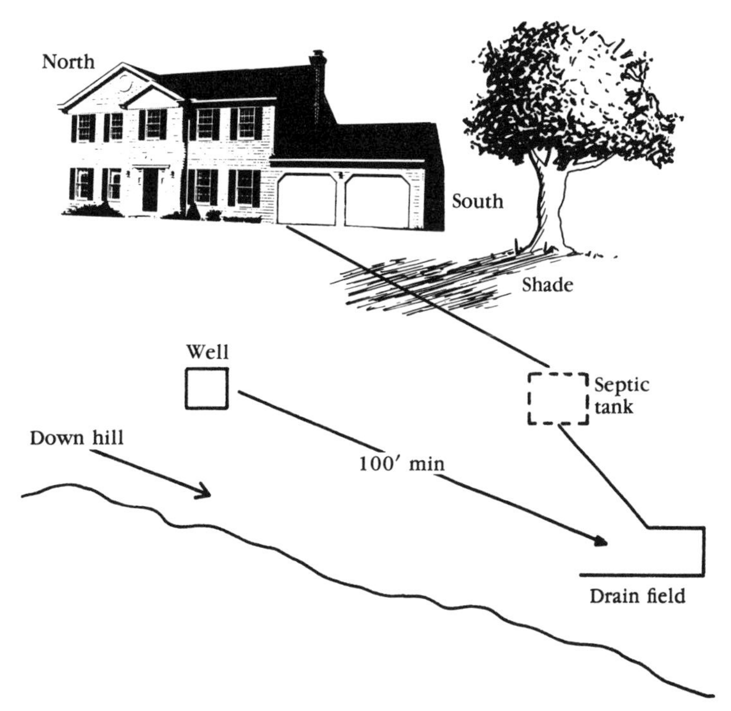

2-1 The area you have chosen for your leach field should be clear of the well, away from the shady side of the house and any shade trees, and on the downhill side of the house.

The basic rules for locating the leach field are:

- The field should be as far away as possible from all water wells (yours and your neighbors'): 100 linear feet is a typical minimum distance.
- The field should be on ground lower than the ground on which any water wells are situated.
- The spot selected should not be a natural sump where rain collects and remains.
- The field should be positioned on the sunny side of the building or property.
- The field should not be close to trees or shrubs, as roots can easily clog and destroy drain field pipes (root systems are about equal in size and extent to branch systems). Trees and dense shrubs also cast heavy shadows that slow evaporation and tend to keep the soil damp and cool enough to freeze sooner in cold weather.
- The earth in which the drain pipes will be buried should be highly permeable.

Highly permeable earth

The ideal drain field soil is light, airy, porous, and dry. Unfortunately, few septic system site soils have these properties, and you must ascertain what sort of soil you actually do have. If there is any choice for field location, you must also determine which area contains the best soil for leaching purposes.

Neither determination can be made by examining the ground surface. You must dig several test holes; you cannot depend upon what just one or two will disclose.

The earth is naturally stratified. If you are on the bed of an ancient river, for example, digging at spots located a few feet in any direction might uncover different soils, washed up eons ago by a long-forgotten river. If there has been extensive building in your area, you might not uncover virgin soil, but you might dig into fill dumped there during land reclamation. A single hole will not reveal the nature of the soil throughout the entire area.

Another factor to consider is depth. The soil near the surface might be heavy with undesirable clay, but there could be highly desirable sand just below. On the other hand, conditions could be the reverse. The surface layer might be sandy, with solid clay or hardpan immediately below. You must be reasonably certain that your site is appropriate, and that means digging at least several test holes deeper than the planned trench depth.

What to look for

The best soil is pure sand and the absolute worst is no soil at all—consolidated rock. Solid clay is not much farther down the unwanted list. What about in-between soils? How do you determine when a particular soil falls between the two extremes?

If you have a soil conservation office (SCO) in your area, you could bring a sample to their office and have its composition tested. Your local building or health department might already have made this test; if not, they should be able to direct you to the nearest SCO. Take care to sample the earth at the depth you plan to set your drain pipes. Make certain your sample is representative of the entire area, not just a random pocket of sand.

You can make a crude estimate of soil drainage properties by the feel of the soil in your hand. Pure clay can be worked almost into the form of a toothpick and it will still keep its shape. Sand, even when soaking wet, can hardly hold a ball shape without collapsing.

Try to shape a sample of damp soil in your hands. The thinner you can work or shape it the greater the percentage of clay that is present. When the soil will only hold a large, crude shape, you know considerable sand is present. The degree of plasticity exhibited by the soil sample between these two extremes indicates the clay-to-sand ratio.

There are exceptions. If the soil is rich in humus, or decayed organic material, it will not be very loose, yet it will be low in clay content. However, humus is generally limited to the top few inches (topsoil). It is dark and easy to recognize. Subsoil is usually light colored, and it is in this layer that your leach lines will be placed.

Another exception may occur when the soil is sopping wet, saturated. In this condition the soil is likely to exhibit much greater plasticity than when merely moist.

Another test is made by placing a ball of soil in a pan of water. Pure clay will just lie there. Puddling or swirling the water will eventually cause the ball to disintegrate. The sandier the clay, the faster the ball will come apart. Pure clay, when mixed with as much water as it will absorb, forms a slippery, shiny, solid mud. This clay can be formed into a cup that will safely carry water, even without being fired into stoneware.

A positive way to determine soil composition is to contract with a soils engineering firm to take and analyze core-drilled samples. They will drill the soil at numerous test sites in the potential leaching areas, make their assessments, and furnish you with a report. Soils are generally classified as shown in TABLE 2-1, and permeability depends upon the exact mix. You may be able to determine the approximate drainage capability of your soil yourself, using this table, if the soil constituents are well defined.

PERCOLATION TEST

The purpose of the percolation test is to determine by actual experiment the rate at which the soil in your proposed drain field can absorb water. This is done by repeatedly pouring water into one or more holes in the ground and timing the rate at which the water disappears. There are various procedures and variations thereon for making this test, which often depend upon the requirements of the local testing or inspecting authority. Details and requirements vary considerably from place to place, but the following is a general outline of the process.

Test holes

If the plan is to run the leach pipe in a straight line along fairly level ground, one hole may be deemed sufficient, but three to five will make a more definitive test. If the leach line is to be laid in more or less a spiral, or a main feed with several parallel branches laid over a roughly square or rectangular area, three test holes should be the minimum.

The reason for having several test holes is to ensure that you are determining the average percolation rate of the entire leach field (FIG. 2-2), not just a small part of it or a chance pocket of clay (poor) or sand (excellent) that would give a misleading result.

Hole depth

The depth of the leach line below grade depends upon climate (and local regulations). The warmer the climate, the closer to the ground surface you can place the drain pipe. If there is any chance of anything

2-1. Soil Classifications and Properties

Soil group	*Unified soil classification symbol*	*Soil description*	*Allowable bearing in pounds/square foot with medium compaction or stiffness*[1]	*Drainage characteristics*[2]	*Frost heave potential*	*Volume change potential expansion*[3]
Group I excellent	GS	Well-graded gravels, gravel-sand mixtures, little or no fines.	8000	good	low	low
	GP	Poorly graded gravels or gravel-sand mixtures, little or no fines.	8000	good	low	low
	SW	Well-graded sands, gravelly sands, little or no fines.	6000	good	low	low
	SP	Poorly graded sands or gravelly sands, little or no fines.	5000	good	low	low
	GM	Silty gravels, gravel-sand-silt mixtures.	4000	good	medium	low
	SM	Silty sand, sand-silt mixtures.	4000	good	medium	low
Group II fair to good	GC	Clayey gravels, gravel-sand-clay mixtures.	4000	medium	medium	low
	SC	Clayey sands, sand-clay mixtures.	4000	medium	medium	low

	ML	Inorganic silts and very fine sands, rock flour, silty or clayey fine sands or clayey silts with slight plasticity.	2000	medium	high	low
	CL	Inorganic clays of low to medium plasticity, gravelly clays, sandy clays, silty clays, lean clays.	2000	medium	medium	medium[4]
Group III poor	CH	Inorganic clays of high plasticity, fat clays.	2000	poor	medium	high[4]
	MH	Inorganic silts, micaceous or diatomaceous fine sandy or silty soils, elastic silts.	2000	poor	high	high
Group IV unsatisfactory	OL	Organic silts and organic silty clays of low plasticity	400	poor	medium	medium
	OH	Organic clays of medium to high plasticity, organic silts.	0	unsatisfactory	medium	high
	Pt	Peat and other highly organic soils.	0	unsatisfactory	medium	high

[1]Allowable bearing value may be increased 25 percent for very compact, coarse grained gravelly or sandy soils, or very stiff fine-grained clayey or silty soils. Allowable bearing value shall be decreased 25 percent for loose, coarse-grained gravelly or sandy soils, or soft, fine-grained clayey or silty soils.

[2]The percolation rate for good drainage is more than 4 inches per hour, medium drainage is 2 to 4 inches per hour, and poor is less than 2 inches per hour.

[3]For expansive soils, contact local soils engineer for verification of design assumptions.

[4]Dangerous expansion might occur if these soil types are dry but subject to future wetting.

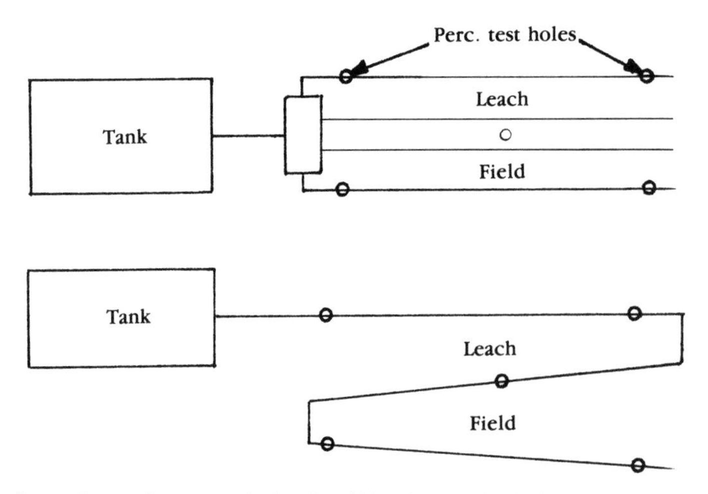

2-2 Several percolation test holes should be dug to check the whole leach field area.

heavier than people passing over the pipes, they should be at least one foot deep, measured to the top of the pipe. Where there is frost, the drain pipe must be deep enough to preclude the effluent freezing. The frost line depth used by builders for foundation depths, however, does not apply here. Effluent is not pure water, and its freezing point is substantially lower than that of water. The safe depth below grade most commonly used for drain lines is 18 to 24 inches.

Check with your local governing authority to learn their recommended drain field depth. You can also check with neighbors to learn their leach field depth and winter experiences. Aerobic bacteria work best when they have maximum air. Obviously, the deeper you plant the leach lines the less air the little creatures will have. The shallower the lines can be placed, the better, and the less effort that will be involved as well.

Making the holes

To make the test holes, dig pits, with relatively flat bottoms, to a depth of 3 to 4 inches below the planned depth of the bottom of the leach pipes. Depending upon the required depth and the nature of the soil, and especially if the holes are being dug by machine, you may have to dig a sizable hole to work in, then hand-dig the actual test hole in the bottom center of each pit. Make the test holes themselves vertical-sided, 8 to 12 inches across, round or square, and about 14 inches deep (FIG. 2-3). Roughen the side walls with a wire brush or other tool to break up any surface glaze or compaction that might affect the test results. Scoop out any loose dirt and place about 2 inches of clean sand or gravel on the bottom of the hole, and level it.

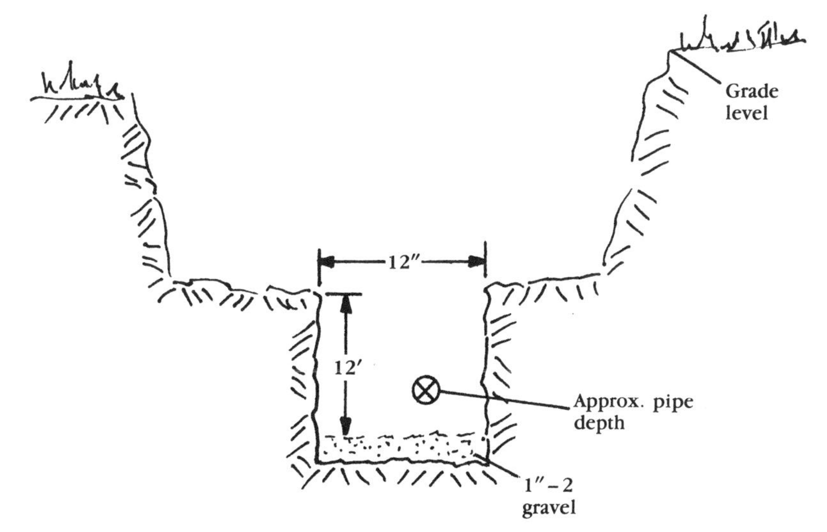

2-3 Perc test holes are typically structured like this.

Conducting the Test

Fill each hole with about 12 inches of clean water, taking care not to disturb the gravel bottom or knock soil off the sides. Keep the holes fairly well filled to that level for at least 4 hours (some authorities require 8 hours, or overnight).

Finally, pick a convenient starting time, such as right on the hour. Adjust the water level in each hole to exactly 6 inches deep. You can check this level easily enough with a ruler or a yardstick bottomed against a small flat rock set in the center of each hole (FIG. 2-4). Wait for exactly 30 minutes, then carefully measure the water level in each hole and make a note of the depth. Multiply the drop in level to the nearest ¼ inch by 2. Add the numbers for all the test holes and average them by dividing the sum by the number of holes. The result is the percolation rate, expressed in terms of inches of drop in water level per hour. (The rate is also sometimes calculated in minutes per inch of drop.) Note: If you are testing at several holes, starting the timer for each hole 5 or 10 minutes apart makes the job easier. If you have any question as to the accuracy of the test, refill the holes and run another series (this step may be required anyway).

Sometimes the tests don't work out quite as just described. For example, if water seeps into the test holes as you dig them, you are out of luck—there will be no soil absorption there. Also, if the rate is less than 1 inch per hour, the soil is unacceptable for any kind of soil-absorption septic system.

If the test holes empty themselves overnight, or in less than 8 hours, you are in luck. A slightly different test procedure is then commonly

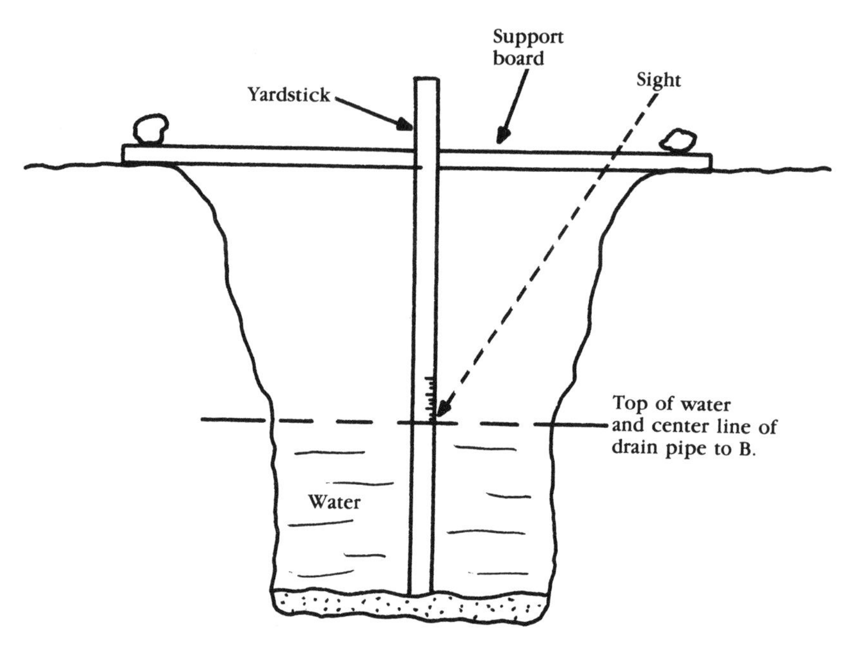

2-4 This is one easy way to measure the water level in a perc test hole.

used. Keep the test holes filled to a 6-inch depth for 3½ hours. At 3 hours 30 minutes, fill the holes exactly to 6 inches deep. At 4 hours measure the drop, and multiply by 2 to get the perc rate.

If the water disappears practically as fast as you can keep up with it, try this. Hold the water level to 6 inches deep for 50 minutes. At the end of the 50 minutes, set the level to exactly 6 inches and start timing. At the end of 10 minutes measure the level drop and multiply by 6 for the perc rate.

Be sure to keep notes of all your readings, along with a layout sketch of the test hole locations. You will need this data to determine the amount of absorption area required in the leach field, as well as its layout. Usually, a perc rate of 1 inch per hour, although marginal, is enough to support a properly designed soil-absorption septic system, and 2 inches or more will handle any of several designs. The greater the rate, the smaller the leach field area need be. Local regulations vary widely on this score, as well as on field design.

Chapter 3

Selecting a tank

The purpose of the septic tank is to still the flow of sewage from the building and contain it long enough for anaerobic bacteria to digest the fecal and other organic matter. This period is estimated to be from three to five days. The tank must therefore be at least large enough to contain three days' worth of sewage. An oversized tank does no harm; and undersized one fills too quickly and requires more frequent cleaning out.

If the size of a leach field matches an undersized tank and the tank becomes overloaded, raw sewage is likely to flow into the field. Black scum might appear on top of the leach field, consisting of anaerobic bacteria and undigested sewage, with a most unpleasant odor. Pools of water might also appear because the undersized leach field will not permit proper drainage, evaporation, and transpiration.

Because the tank contains a tremendous quantity of pathogens, it must be watertight at all points below the working liquid level. It must be impossible for any of the contained liquid to leak into the soil surrounding the tank.

Because the action of anaerobic bacteria produces an odor and several gasses (mainly methane, which is flammable), the tank itself must be nearly airtight. To prevent the tank from bursting due to the pressure of gasses being constantly generated by the anaerobic bacteria, the tank must be vented.

Because you don't want to replace a tank every few years, it must be made of some rot- and rust-resisting material. The most practical place for the tank is underground, so it must be strong enough to support a layer of earth, as well as any traffic that may pass over it.

SELECTING THE MATERIAL

Tanks are commonly made from several materials. Your chore is to find the most satisfactory compromise between tank affordability (installed) and life span. Costs depend on whether or not you make the tank yourself or purchase one ready-made, kind of material selected, and distance from the supplier. The weight of the tank needs to be considered. A precast concrete tank, for example, requires special equipment for lifting and placing, but two men can manually handle a comparably-sized fiberglass tank.

Redwood

A redwood tank has an average life span of 30 years. It can be constructed on-site by the homeowner. It should be coated on the inside with a bituminous waterproofer. It rots most quickly at the waterline, from the outside in. It cannot be set too deeply in the earth because the dirt load on top can be a problem and bracing is expensive.

Fiberglass

The life span of fiberglass tanks is estimated to be 30 years, but no one knows for certain because the product has not been around long enough. Fiberglass tanks consist of glass fiber cloth or mat bonded with plastic materials. Because all the ingredients are nonbiodegradable, the tank's actual life may be much longer than is presently estimated.

Fiberglass tanks are made in many sizes, and a unit large enough for an average home weighs no more than 300 pounds. Two men can position a tank in the excavation. Make certain the tank is strong enough to withstand the soil load at its intended depth.

Metal

The most practical metal from the standpoint of cost is mild steel. The most practical method of fabrication is by electric arc welding. Both inside and outside surfaces should be coated with a bituminous waterproofing material. A tank, made in a shop and hauled to the site on a truck could be lowered into the excavation by a crane, but a backhoe or front loader is more practical. Once commercially available everywhere, steel tanks are less common today. The estimated life of a steel septic tank is 7 to 10 years, but in fact many last for 20 or more years.

Cast concrete

A cast concrete tank is made by erecting a suitable form and filling it with concrete. Depending upon the size and shape of the tank and its cover, one or both may contain reinforcing mesh or bars. The life span given for concrete tanks is 20 years, but a well-made tank will actually have an almost indefinite life span, barring accidental damage or structural damage not caused by normal use. The inside of the tank should be coated with a bituminous waterproofing material.

Concrete block

There are two types of concrete block: standard, made up of cement and gravel aggregate; and lightweight, which contains a light aggregate like cinders, pumice, or vermiculite. Lightweight block should never be used for septic tank walls. After the blocks have been laid up, the interior should be given two ⅜-inch thick coats of cement plaster, allowing curing time between coats. After the final coat has cured, a bituminous waterproofing material should be applied.

The life span of a concrete block tank is generally assessed at 20 years, but if well plastered and waterproofed and if no mechanical damage occurs, it should last considerably longer.

Brick

The life span of a brick tank may be a bit less than that of a block tank, but much depends upon the kind and quality of brick used. Useful life can be substantially prolonged by cement plastering and waterproofing the inside walls.

SIZING THE TANK

The septic tank must be large enough to contain a 3- to 5-day quantity of sewage. Whether your tank holds 3 or 5 days sewage is a difference of 66 percent, and makes a correspondingly large difference in tank size, cost, and labor. Unfortunately there are no firm guidelines for favoring one time period over the other. However, temperature conditions at the site should influence the final design, as bacteria work much more slowly as temperatures lower. The colder the weather at the site, the larger the tank should be.

Oversized tanks

An oversized tank has other advantages. The effluent will remain longer in the tank, resulting in a greater degree of bacterial digestion and breakdown of solids. Consequently, the liquid entering the drain field will have a smaller percentage of solids. Solids in the effluent entering the leach field pipes are a major factor in eventually clogging the field and forcing you to renew it. Thus, the "cleaner" the effluent entering the leach pipes, the greater the service life of the field.

There is another advantage. The solids that anaerobic bacteria cannot digest remain at the bottom of the tank. Eventually, so much solid matter can accumulate that tank capacity becomes reduced. In the worst case, solids build high enough to block the inflow of sewage, or cause solids to flow out in the effluent, or both. The tank must be pumped out before this can occur. Typically this is a routine maintenance chore that should be performed about every 2 to 3 years depending upon tank size, location, and service conditions. An oversized tank will save the homeowner money by reducing the frequency of required cleanouts.

Undersized tanks

The initial cost of an undersized tank will be less than that of a similar, but larger, tank. But undersize, unfortunately, has no definite measurements. You can only be certain that a tank is undersized when operating problems result. To make matters worse, some of these problems may not surface for years. If, for example, a properly sized and installed leach field clogs up in 10 years instead of 20, the premature failure likely can be ascribed to an undersized tank.

If a septic tank has to be pumped out and cleaned of solids every 6 months rather than every couple of years, it's a certainty you have an undersized tank. If the earth above the leach field is frequently wet, if it emits an odor, if black muck appears on the surface, tank undersizing is most likely the cause.

Codes

Many municipalities have building codes that specify the size of the septic tank on the basis of either the number of bedrooms or the potential number of occupants in the building. No matter what method local authorities use for sizing a septic tank, the homeowner must comply. But nothing prevents your selection of a tank larger than the prescribed minimum. Some building departments also specify the tank type, manufacture, position, depth, pipe arrangement, and other details as well. Don't try to circumvent the building department. Though you might not agree, the regulations generally benefit the homeowner and the community.

Bedroom versus head count

The old method of determining required septic tank capacity was by counting bedrooms. This is unreliable; a bedroom can sleep one to half a dozen or more. A two-bedroom house might be converted in time to a four-bedroom house. A more accurate way of estimating the quantity of water that will flow out of a house is to count the number of people living permanently in that house. Of course this, too, is problematical, because the number of people living in a house changes from time to time. A logical assessment must be made, according to the house size and potential occupancy. Err on the high side, rather than low.

The usual procedure is to count all people over the age of five, since children must also bathe, shower, eat, and have their clothes washed. In fact, it does no harm and may be wise to include younger children as well, because a considerable amount of waste water is accumulated on their behalf. And if you want to be prepared for the onslaught of time, you need to assume that children may use more water as they grow up, and that more people may join the family.

Realistic estimate

Chapter 9 explains how to estimate the total quantity of water the average individual uses per day: the figure is 95 gallons. This includes all water used for cooking, washing, and toilet flushing.

Assuming five people live in the house, water use would then amount to:

$5 \times 95 = 475$ gallons

Three days would bring the total to:

$3 \times 475 = 1425$ gallons

Five days would bring the total to:

$5 \times 95 = 2375$ gallons

When you multiply it all out, the two figures are considerably different. Which to use? To be certain, going with the larger-volume tank is best. On the other hand, if you live in a warm climate and have no reason to believe your family will grow larger, going with the lower figure or making an in-between compromise is reasonable.

REDUCING THE TANK SIZE

There is a practical way to reduce the required size of a septic tank and its associated leach field without reducing either efficiency or safety of the system. The method is not permitted in all areas of the country—be sure to check with local building or health officials.

Two kinds of water flow through the main drain of your house. The water that comes from the toilet is called black water. That coming from the tub, shower, laundry, and washbasins is called gray water. Black water must be treated by anaerobic and aerobic bacteria and filtration before disposal. Gray water can be more directly disposed of by one of two methods. Kitchen water, while also gray water, requires slightly different treatment.

Division of waste water

According to the National Water Well Association, water consumption in the American home breaks down this way:

- Bathroom: 75%
  - Toilet flushing: 45%
  - Bathing: 30%
- Kitchen: 25%
  - Dishes and laundry: 20%
  - Cooking and drinking: 5%

About 45 percent of all the water used in the average household goes to flushing the toilets. This is black water. It must go through the septic tank and leach field. Another 50% percent is a combination of dishwater water, bathwater, and shower water. This is classified as gray water. It does not have to be similarly treated because it ordinarily contains no pathogens. Dishwashing water contains oils and greases that should not be mixed with bathwater.

By separating the black water from the total volume of water flowing out of the house sewer pipe and into the septic tank and limiting the flow to black water alone, the required safe size of the tank and leach field can be cut in half (FIG. 3-1). This is accomplished without reducing the efficiency of the tank or the field. Other steps, of course, have to be taken for handling bath and kitchen water.

Some low-flush toilets use as little as 1.5 gallons of water per flush, saving more than 40 percent of the normal amount of water that becomes black water. The Japanese manufacture toilets with two flush systems, one for feces and the other for urine. The quantity of water used for each flush depends upon the direction in which the control handle is swung. Some models have a lavatory atop the toilet tank. When you wash your hands you produce gray water, which then goes into the toilet tank and is used for flushing.

Some homeowners connect their tubs, washing machines, lavatories, showers, etc., to their toilets to use the gray water for flushing. To make sure there is some kind of water always available, and to relieve or divert the gray water flow should the tanks be full, some sort of automatic pressure valve must be installed in the system. In addition, all traps in the drain lines must be readily accessible. Gray water usually contains a quantity of hair which in time clogs the traps and must be removed.

There are also composting toilets that operate without using any water at all. This sometimes allows the septic tank and leach field to be dispensed with entirely, at least in its usual form. The technology used is reasonable and safe, but initial equipment costs tend to be high and public and governmental acceptance has been, unfortunately, slow.

GRAY WATER DISPOSAL

There are two alternative ways to dispose of gray water. One method is legal nearly everywhere, the other is frequently prohibited. The first consists of directing the water underground, the second involves simply dumping it out on the surface of the ground.

Gray water is used by airlines, colored blue and used to flush the toilets. Sud-saver clothes washers retain water used for the first washing and recycle it for the second washing. The Grand Canyon National Park, and perhaps others as well, use gray water for toilet flushing. Campbell Soup saves the water it uses for washing its vegetables and directs it to its fields to irrigate crops.

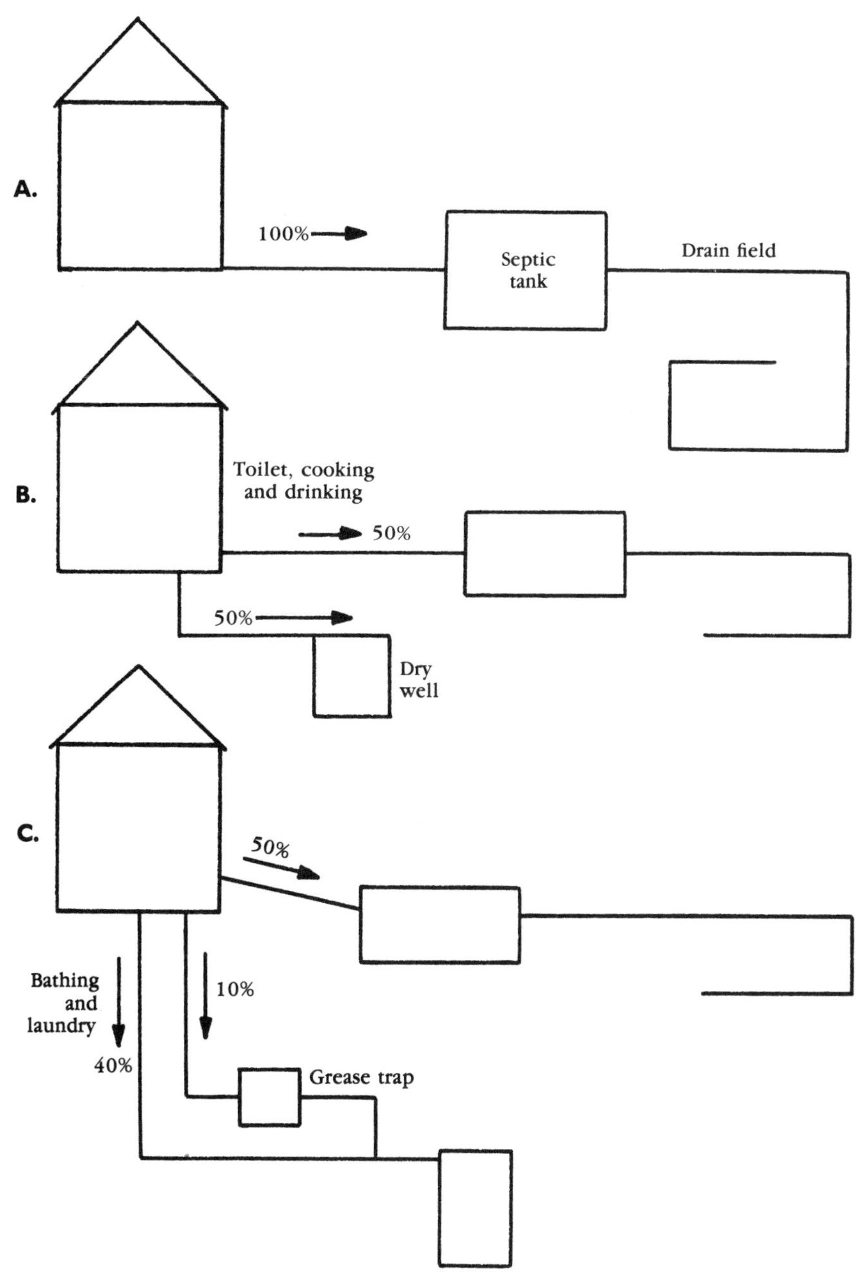

3-1 A. The total flow of sewage out of the house enters the septic tank. B. Here the sewage is divided into black water (50%) and gray water (50%). The gray water is led to a dry well, thus cutting the load on the septic tank and drain field in half. C. To keep the dry well from being clogged with cooking oils and grease, the kitchen water is run through a grease trap before it enters the dry well.

In many parts of the country, gray water can be used to water trees, bushes, and grain crops. Used this way, the gray water does not come in contact with the edible portions of plants. In other words, you may water your lawn with gray water, but you may not use the same gray water to irrigate celery or lettuce. There is a potential danger that pathogens might come in contact with the vegetables and be ingested along with them.

Underground disposal

Very simply, the gray water is led off to a *dry well.* The dry well must be large enough to handle the gray water, and the pipe from the building to the dry well must be well insulated and perhaps wrapped with heat-tape, or buried deep enough in the ground to avoid freezing.

Take 50 percent of your home sewage volume as the gray water volume (and reduce the black water volume by an equal amount). You now have the required volume in gallons per period needed for the dry well, which may or may not equal the required capacity of the well.

The rate at which water will drain from the well into the soil also governs well capacity. For example, if the well is in coarse sand or gravel, liquid will leave the tank and enter the soil faster than it enters the dry well. Under such conditions, you can safely reduce the dry well size. At the other extreme, you might encounter solid clay with a very low rate of absorption. In that case, you would have to construct a substantially oversized dry well and probably alter the composition of the adjacent soil by adding sand or gravel.

Chapter 2 describes techniques that can be used to measure the porousness of the soil and estimate the rate at which it will absorb water.

To size a dry well, take half your total estimated figure of the quantity of water that will be used by the householders. For example, look at a home with five occupants. Each individual will use 95 gallons of water per day, a total flow of 475 gallons a day. Half, or 237 gallons, will be gray water.

Increase that figure to 250 gallons per day for convenience and to provide a small safety margin. You will need a dry well with a total liquid capacity of 250 gallons (minimum). To convert this figure into volume in cubic feet, multiply 250 by 0.1337 = 33.4 cubic feet.

To find the dimensions of a tank that will contain 33.4 cubic feet or 250 gallons of liquid, the easiest way is to find the cube root of the volume, which equals approximately 3.4 feet. Thus, a cube-shaped tank with inside dimensions of 3.4 × 3.4 × 3.4 feet will hold about 250 gallons. This appears to be the required size of the tank, but hold on. You are not out of the well yet.

Projecting that your need will be satisfied by a 33.4 cubic-foot tank assumes the tank is empty until completely filled with gray water, and the water will be completely absorbed by the surrounding soil at a rate of one full tank per day. Neither assumption may hold true.

Types of dry wells

There are three principal dry well designs, called tank, positioned block, and rock pile. The tank design provides 100 percent volumetric efficiency. No interior space is lost. The positioned block design is about 75 percent efficient and the rock pile about 33 percent. If you construct a tank, you can build to a convenient inner dimension of 3.5 × 3.5 × 3.5 feet to secure the 250-gallon volume needed. With positioned block, you must increase the tank's inner volume by at least 125 percent. If you select the rock pile, you have to at least triple the volume.

Besides considering the type of dry well constructed, the soil absorption rate should be measured, and the resulting figure used to determine the necessary exterior area of the tank, less its top. Only the sides, usually below the inlet line, and the bottom are considered absorption areas.

Building a tank well

Tank wells are true tanks—they are completely empty and have perforated masonry walls. The floor of the excavation serves as the tank bottom.

Excavate the required hole, making the sides as vertical as is practical. Level the bottom and cover it with a 4- to 6-inch layer of gravel, leveled. Lay the walls up with standard 8-inch concrete or cinder blocks spaced apart ¼ to ½ inch. Spread a shell-bed of mortar between the courses, but leave the end joints dry to provide drainage gaps.

Build the walls straight up until you approach the last 2 or 3 feet, and then begin to corbel the blocks inward. This reduces the size of the top opening. Using mortar, plaster the entire exterior of the corbeled section to strengthen it; treat the interior similarly if you wish. As you work your way up, use baffles or other means to keep fresh mortar from slipping into the joints between adjacent blocks.

Terminate the well top at least a foot below finish grade. All the joints in the last course should be mortared, the top cores filled with mortar, and the exterior surfaces plastered with mortar. This will keep the upper blocks from absorbing excessive surface ground water. Set a lid of 3-inch bluestone or approximate equivalent over the opening; a cast concrete lid is another possibility. Backfill the excavation with ¾-inch gravel. Figure 3-2 shows the details.

Positioned block

Use standard 8-inch or 10-inch cinder or pumice blocks. Place the blocks, with cores set vertically, in an approximate circle within the excavation. Loosely fill the spaces with more blocks. Position the second course of block atop the first, leaving plenty of gaps. The object is to fill the entire hole using as few blocks as possible, retain the dirt walls of

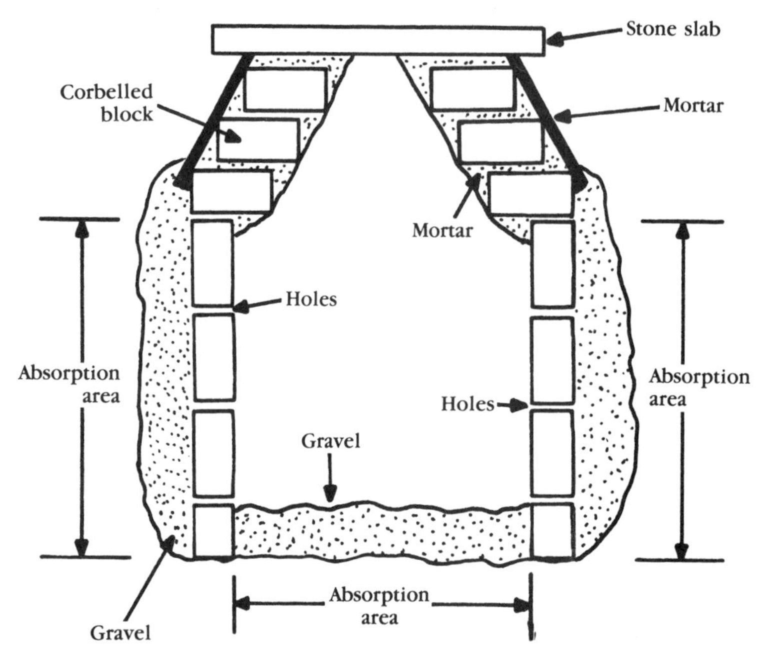

3-2 One design for a tank type of dry well. The upper courses are corbeled to make a small stone slab cover possible.

the excavation, and still provide a stable support for a stone or concrete slab cap that will support the earth roof.

You will use at least twice as many blocks with this design as with the tank design, but you can use damaged or used blocks and chunks and not be troubled with mortar and actual masonry work. For a small well, this method is practical. As the well size increases, however, the ratio of block quantity to well size rapidly becomes too problematical and expensive.

For best results, the space between the blocks and the wall of the excavation should be filled with coarse gravel, but this must be done with care to avoid disrupting the block arrangement too much. Top the block with flat stones and tar paper.

Constructing the rock pile

Excavate a hole of suitable size, and fill it with loosely piled rocks. Quarry-run rock, leftover material from a stone quarry, is the least expensive stone commercially available, but is just what you want—ill-assorted, odd-shaped, and varied in size. They are just the kind of stones that will not pack evenly and will form large voids when dumped into the hole. Random rubble stone and glacial till fieldstone are other possibilities. If you scrounge for stones, go for the larger sizes (bowling-ball and up) with rough surfaces and irregular shapes. When you dump them into the hole, try to keep them from packing or nesting together. The 33-percent efficiency for this design is derived from the fact that odd-shaped stones in a pile usually have voids between them that account for one-third the total volume of the pile.

To complete the rock pile well, place flat rocks and gravel on top of the topmost layer of stone to provide a more or less flat surface. Cover the surface with a double layer of tar paper, then backfill with sufficient soil to form a low mound over the well. The mound will eventually compact and settle out level. See FIG. 3-3.

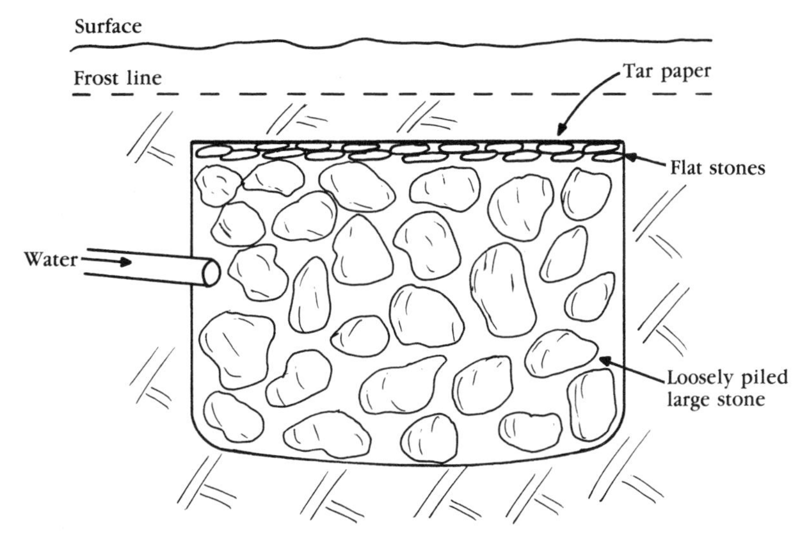

3-3 A rock-pile dry well, where fieldstone or quarry rubble is piled helter-skelter in the hole. Flat stones are saved for the top to provide support for the tar paper that keeps soil out until it settles and compacts.

The tank well is far and away the best choice. It has greater liquid capacity for its size than the others, and can be cleaned with a reasonable amount of labor to restore its efficiency. The other two types must be completely dug up and replaced.

GRAY WATER PRECAUTIONS

Gray water contains phosphorus and nitrogen in various quantities. Few plants can take gray water as a steady diet and thrive, so alternate plant-watering with fresh water. Don't water spinach, cabbage, lettuce, or other vegetables or fruits at all with gray water if the water will come into direct contact with the edible portions. Use gray water for washing the car, hosing off the driveway and walks, and similar chores.

If you are running the gray water through a hose and onto your lawn or garden, cover the end of the hose loosely with a cotton bag to filter the water. Change or wash the bag often.

TREATING KITCHEN WATER

Most homeowners do not attempt to separate the kitchen sink and dishwasher drain from the black water. This is a double mistake. The water coming from these sources contains oil and grease, not easily digested by the anaerobic bacteria in the septic tank. Most of the oil and grease floats to the surface of the liquid in the tank and forms a scum. In time this will combine with laundry and dishwashing detergents and enter the leach field, soon clogging the soil. While this practice does not induce overnight leach field failure, it certainly does shorten service life.

Grease trap

The proper treatment for the kitchen sink and dishwasher drain water consists of running it through a grease trap (FIG. 3-4). Traps can be purchased ready-made from a plumbing supply house or fabricated from concrete or metal. A trap is a 1- to 2-cubic-foot container with one or more baffles and a watertight, removable cover. Greasy gray water flows in one end, is slowed by the baffles, and flows out again.

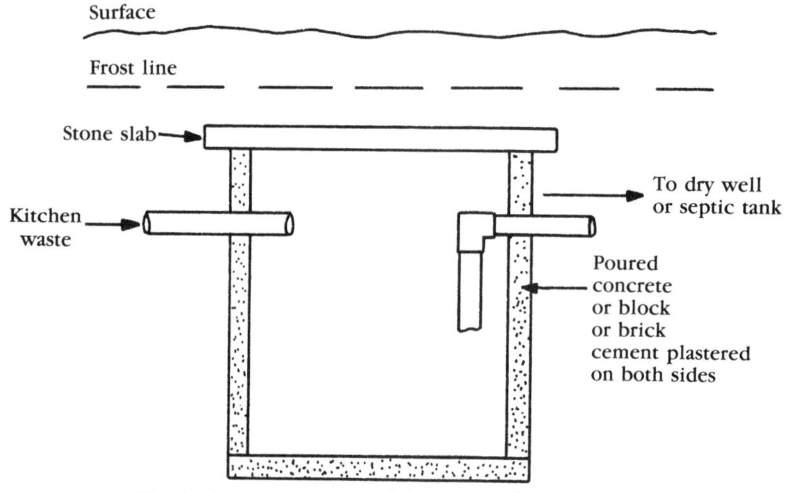

3-4 The basic arrangement of the parts of a grease trap.

During the flow to the trap and while it remains inside the trap, the water cools. The grease and oil settle out or float to the surface, and cling to all interior surfaces of the trap. This is why metal can be used for a trap—the grease coating protects it from rust and corrosion. The gray water leaving the trap is essentially free of oils and grease.

The trap should be located in an accessible spot that is cool but not cold. You want the liquid to be chilled but never to freeze. Every year or so the trap should be inspected and the grease scraped out and discarded.

To save on piping it is possible to install a grease trap in the house main drain, where both gray and black water will pass through it. This method works, but cleanout is an unpleasant task at best, and the trap is a potential source for drain clogging and consequent sewage backup.

SEPTIC TANK DIMENSIONS

You have counted heads and decided on the capacity, in gallons, of the septic tank. Now convert the gallons to cubic feet. A cubic foot contains 7.48 gallons. Dividing the required gallon capacity of your tank by that figure gives you the required tank volume in cubic feet.

Because a cube is not the most practical or efficient shape for a septic tank, you can't approximate its size just by finding the cube root of the volume and adding a couple of feet of height to provide the necessary clearance for pipes, baffles, and venting above the working liquid level. The easy way to estimate dimensions is to use TABLE 3-1. Those figures provide for tanks about 25 percent over capacity. No tank *must* be made to the exact dimensions given—they can be varied, and local regulations may require other minimum sizes.

3-1. Minimum Recommended Septic Tank Dimensions

Number of people	*Approx. gallons*	*Inside width*	*Inside length*	*Liquid depth*	*Total depth*
4	250	3′ 6″	3′ 0″	4′ 0″	5′ 0″
6	360	4′ 0″	4′ 9″	4′ 0″	5′ 0″
8	480	4′ 0″	5′ 0″	4′ 0″	5′ 0″
10	600	4′ 0″	6′ 0″	4′ 0″	5′ 0″
12	720	4′ 0″	7′ 3″	4′ 0″	5′ 0″
14	840	4′ 0″	7′ 9″	4′ 6″	5′ 6″
16	960	4′ 0″	8′ 9″	4′ 6″	5′ 6″

Recommendations based on a figure of 95 gallons per day per person, with 40% of the effluent being directed to a dry well (57 gallons of black water into the tank per person per day.) Note: If a garbage disposal unit is to be installed, tank capacity should be increased 50% if the kitchen sink drains into the septic tank.

DESIGNING THE SEPTIC TANK

If you have decided to build your own tank rather than purchase a ready-made one, your next step is to decide on the type of tank to build. The five to choose from are the single compartment, double compartment, triple compartment, meander, and siphon. As you go down the list, the tanks become more complicated, somewhat more expensive to construct, a little more difficult to maintain, but more effective in removing solids from the effluent. The siphon type also provides a more forceful flow of effluent from the tank into the leach field.

The single tank has one trap door and access opening. When it needs to be cleaned, the pump crew opens the trap and loosens the sludge at the bottom of the tank with a long-handled shovel. A large-diameter suction hose is then inserted into the tank and an external pump transfers most of the sludge and effluent to a tank truck. In a double-compartment tank this operation must be repeated for the second section.

The cleaning process must be also done for each section in the three-compartment tank. In the meander tank the number of cleaning operations depends upon its construction, but three cleanout ports are typical and there may be four or more. The siphon design usually contains three compartments, requiring three separate cleanout operations.

Single-compartment tank

Figure 3-5 shows the basic parts of a single-compartment septic tank. The diameter of both the inlet and outlet lines should be the same, and should be a minimum of 4 inches.

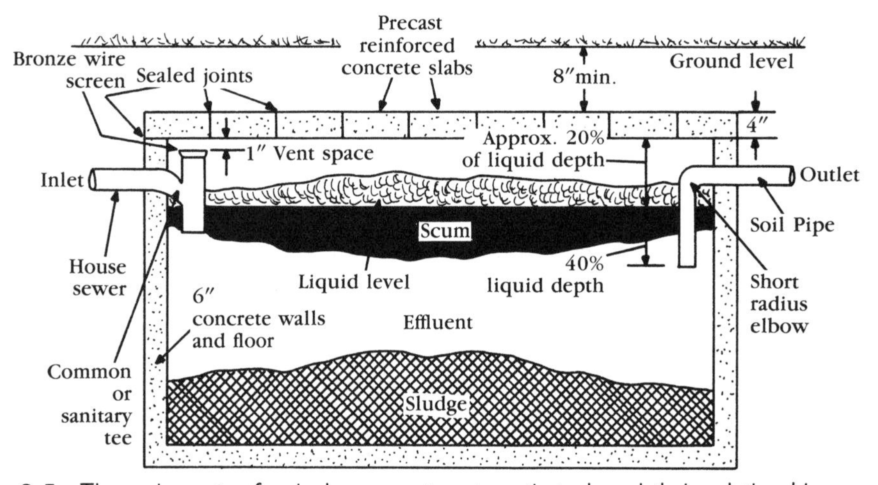

3-5 The main parts of a single-compartment septic tank and their relationship.

Use a tee pipe on the input side and cap the top end of the tee with bronze screening. The top of the tee should be 2 inches below the top of the tank. This construction permits gasses produced by the anaerobic bacteria in the tank to escape via the sewer pipe and fresh-air vent system.

The outlet pipe can be either an elbow or another tee. It should be positioned on a plane with the inlet pipe. If the outlet is lower than the inlet the maximum practical depth of the tank will not be used because the liquid will drain as soon as its level is above the outlet pipe. If the outlet is higher than the inlet, the liquid in the tank will rise above the inlet pipe and create an undesirable back pressure.

Use either vitrified clay pipe for the inlet and outlet or plastic pipe with sanitary fittings, which have no inner ribs. This virtually eliminates the possibility of muck catching and remaining there, and clogging a pipe.

Two-compartment design

In the two-compartment design (FIG. 3-6) the total volume remains the same as a single-compartment tank. Inlet and outlet pipes are also positioned the same. The only change is a divider containing two ports. The divider is placed so as to make the outlet compartment about 40 percent as large as the inlet compartment, and it need not be watertight.

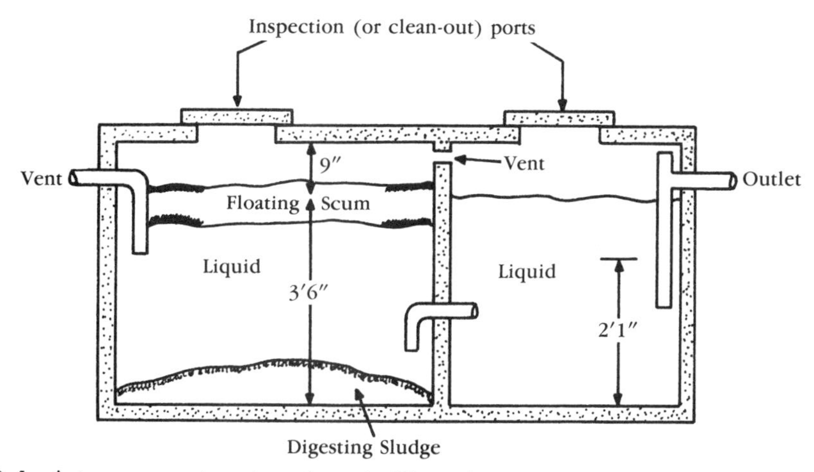

3-6 A two-compartment septic tank. Dimensions are given only to illustrate the approximate relationships of the parts.

In a masonry tank the divider can be fashioned of concrete partition block (typically 4 inches thick) laid on edge. In a wooden tank it can be a wood partition. You can use either a short length of pipe to connect the two compartments, or you can leave an opening approximately 5 inches square in the divider at about the three-quarter height.

The purpose of the divider is to slow the movement of liquid even more than is usual when it enters a single tank. Stilling the liquid is crucial to efficient tank operation; slowing inflow improves this situation. It also allows the microbes more time to act and the solids more time to settle. And, more solids are kept out of the leach field, thereby lengthening its service life.

Because there are two compartments and no easy way of reaching one from the other, two access hatches are needed. The divider will keep most of the solids out of the outlet compartment, but sooner or later the sludge will have to be loosened and pumped out.

Another feature you might find advisable to add is a *relief vent* in the divider. This is a 5-inch-wide, 2-inch-high hole centered at the very top of the divider. Through it gasses travel from one section to the other and vent freely, and it enables similar liquid movement should the lower connecting pipe become clogged.

Three-compartment design

In the three-compartment design, the requirements for the placement of the inlet and outlet pipes and their construction remains the same. The compartments can be divided on the basis of 2:1:1. The ratio need not be exact, but increasing the overall inside dimensions of the tank to make up for the space taken up by the partitions is important.

Again, you need a relief aperture at the top of each divider, and access hatches for each of the three compartments.

The three-compartment tank doesn't do anything different than the one- and two-compartment tanks do, it just does everything better. It keeps the liquid quieter, the effluent clearer, and gives the bacteria even more time to work on the sewage.

The siphon tank

The siphon tank is a bird of a different color. This design can be used with one-, two-, or three-compartment tanks; the original had three compartments (FIG. 3-7). In this arrangement, the sewage flows into the first tank then siphons automatically from the first into the second tank and similarly to the third.

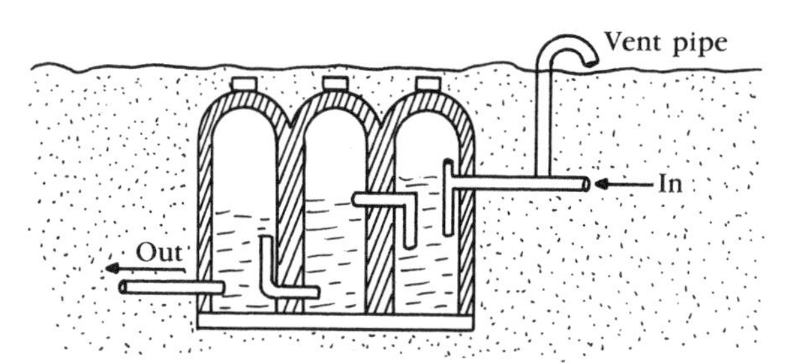

3-7 A three-compartment tank designed in 1883 by Edward S. Philbrick of Boston, Massachusetts. The air vent helps relieve gas pressure in the compartments, but also hinders action of the anaerobic bacteria. The siphon provides a fast-flow discharge from the inflow compartment at the right to the middle compartment.

Figure 3-8 shows the operation. When the liquid level in the first compartment rises above the elbow of the siphon, liquid begins to flow up this pipe and down into compartment two. The siphon pipe is now filled with liquid and flow continues until the liquid level in compartment one falls below the siphon pipe inlet, when siphoning action stops. If there is a third compartment, the same sequence takes place there.

The advantage of this arrangement is pressure. In the previously discussed septic tank designs, the rate of liquid flowing out of the tank is never greater than that of the liquid flowing in. When, for example, a toilet is flushed, there is a momentary gush of perhaps 8 gallons of water. The water, now sewage, enters the septic tank and possibly raises the existing liquid level by a fraction of an inch. A like amount of

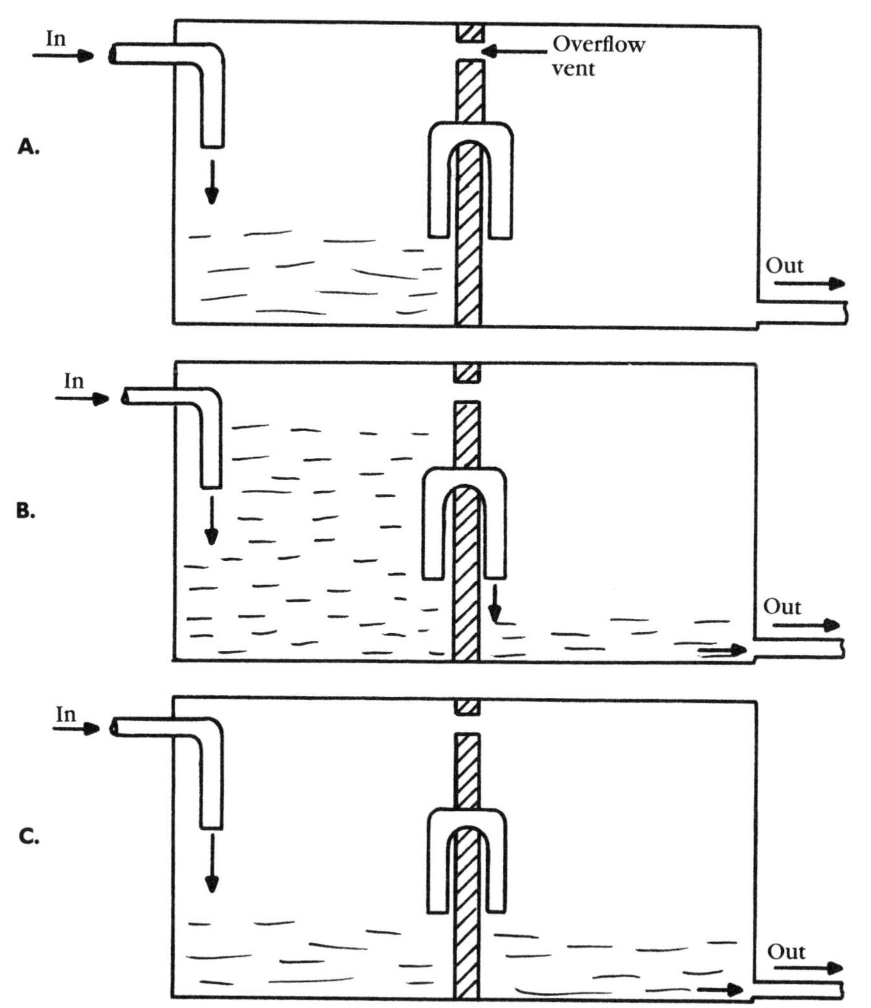

3-8 How a siphon tank works: Black water enters at right. The siphon is empty. The inflow compartment fills to a point above the top of the siphon. Water now flows down the right-hand siphon leg. Water continues to flow down the right siphon leg until the water in the compartment is about level with the water in the outflow compartment. Thus, it is possible to have a succession of half-full levels of effluent in the outflow compartment that will provide far more pressure than the slow entrance of sewage into the inflow compartment.

effluent flows out of the tank, lowering the level again by that same fraction of an inch.

In hydraulic terms, the pressure on the effluent, or head, is less than 0.1 psi (pounds per square inch). The outflow from the tank is therefore a mere trickle. It is not backed by any pressure; it has no appreciable velocity.

In the three-section siphon arrangement, the third tank is normally empty. When the liquid in the first tank rises high enough to start the siphon action, the liquid pours from the first tank into the second at a prodigious rate (a 5- or 6-inch pipe should be used for the siphon). The third tank is quickly filled and emptied, though there are a few moments when it is nearly full.

Assuming that the liquid is 4 feet deep at this moment, and that the effluent is ejected from the lowest possible point in compartment three, it will be propelled by a 4 × 0.43 psi head, or a pressure of 1.72 psi. This pressure is much greater than that possible with any other nonpowered arrangement. At this pressure, although it is comparatively low as water pressures go, the effluent will still be driven with considerable force and velocity. This helps the effluent reach all parts of the leach field and carries along any obstructions present in the leach field feeder pipe.

Unfortunately, the siphon design has one basic limitation. In computing the head that will be developed, the assumption is that the drain pipe will be positioned at the very bottom of the last compartment. But if the outlet has to be positioned at nearly the same level as the inlet, there is no point in installing a siphon; the higher the outlet, the lower the head. Head is directly dependent upon the height of the liquid above the outlet, and is computed at a rate of 0.43 pounds per foot of height.

The meander design

The septic tank design shown in FIG. 3-9, perfected by Dr. T. J. Winneberger, is believed to retain more solids and produce a clearer effluent than the other designs we have discussed.

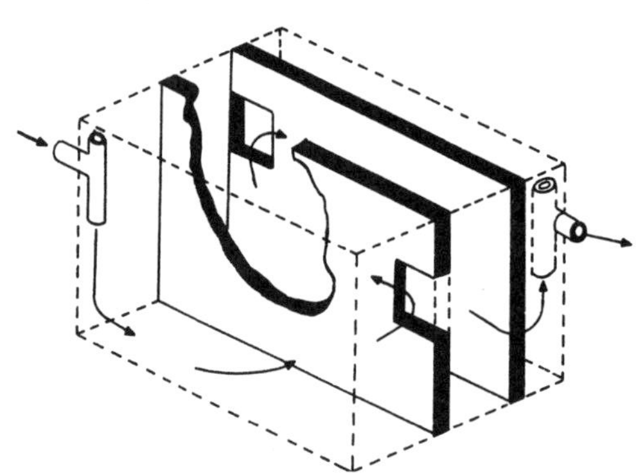

3-9 The meander design forces the sewage to take a long, slow path through the septic tank.

The design was developed to force the effluent to meander like a slow-moving river. The slower a stream moves, the more sand and silt it drops. The more turns a river makes, the more it drops going around the bends. Likewise, by installing partitions lengthwise in the tank and forcing the liquid to travel in comparatively long paths and make several turns, the time the effluent spends in the tank is greatly increased and the more solids settle out.

The first partition has a hole about halfway up its side. The second partition has a hole at the opposite end near its top edge. The outlet is positioned at the far end of the tank. Away from the second hole. Thus the liquid has to travel not only back and forth, but also to some degree up and down. In a tank that is 6 feet long, the path from the inflow pipe to the outflow pipe is about 22 feet.

The meander tank has three compartments, so it must have three access hatches for inspection and cleaning. The partitions can be of any material that will not rust or rot easily. The partitions need not be watertight. In a masonry tank, concrete block can be used for the partitions and the holes are created simply by omitting blocks.

Automatic-valve design

A final design is the automatic-valve or surge tank (FIG. 3-10). This consists of the usual watertight roofed container with a conventional inlet pipe. However, two pipes take care of the outflow. One, for normal operation, extends vertically through the bottom of the tank to a height slightly below the lower end of the inlet pipe, and is open at its upper end.

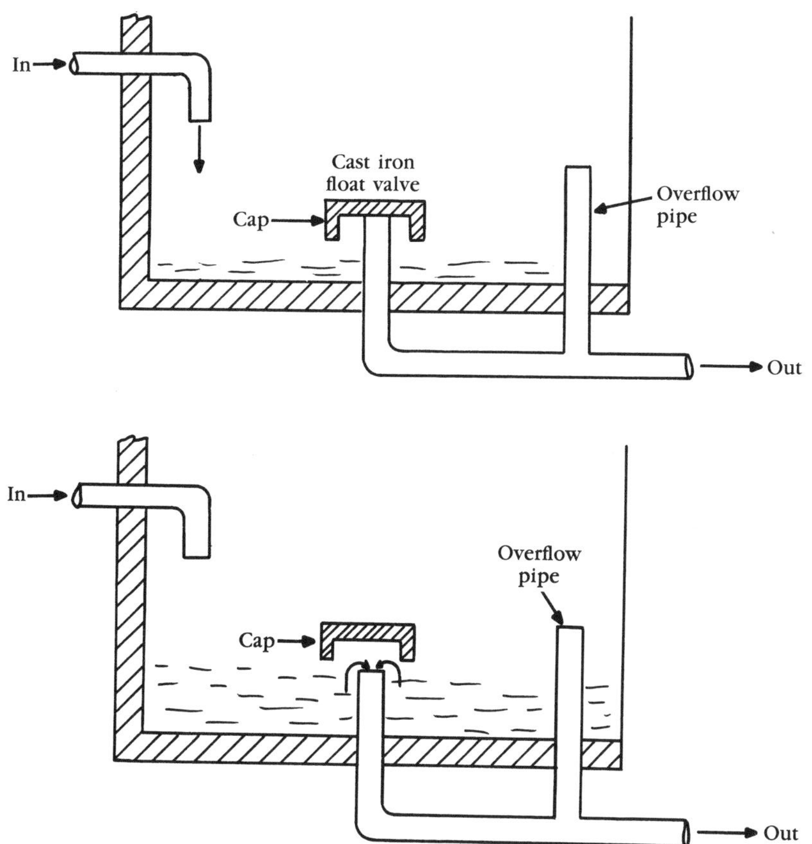

3-10 A simple automatic valve produces some surging action in the outflow of the effluent in this design.

This main outflow line carries a heavy cast iron cap. When the tank is empty or the liquid level is below the cap, the weight of the cap keeps it in place, sealing the pipe. When the liquid level rises above the cap, the air beneath the cap raises it and the liquid can enter the pipe. Thus there is a series of outflow surges interspersed with periods when black water is entering the tank.

The second outflow pipe is also vertical, but extends well above the at-rest position of the top of the cap on the main outflow line. Should the cap become stuck for any reason, the liquid can flow out through this overflow pipe.

Chapter 4

Tank construction

You can purchase a ready-made septic tank and have it set in the excavation you have provided, or hire a contractor for the job. Or, you can hire a contractor to construct a tank in whole or in part right at the job site. To find either a tank supplier or contractor, consult the yellow pages under "Septic Tanks and Systems." You may also have to hire a plumber to make the pipe connections to satisfy building code or union regulations.

Different locales require inspection or approval by the building department or local health authorities of septic tanks and installation. They might insist that a licensed plumber make the pipe connections. Play it safe and check with the local building department before starting construction.

BUILDING YOUR OWN TANK

Building your own tank is sometimes advantageous, especially if you have the expertise and the tank site is inaccessible to a delivery truck, or time is available but cash is not. Irrespective of the tank design you select (refer to Chapter 3) and numerous minor details, the general methods and materials required for tank construction are the same.

No great skill or experience is required to work with any of the materials mentioned, at least in this kind of a job. Neatness isn't important because the tank is out of sight, and precision or fussiness is not required. Squareness and exact dimensions are not crucial either, so long as strength is unaffected. Required tank capacity is at best an estimate. Only the depth of the tank, its orientation with respect to the main house drain, and correct positioning of the inlet and outlet pipes for proper flow, require particular care. Other than that, just make it strong and watertight.

Note: We will not describe wood or brick tank construction in this book. Neither are encouraged and in some areas are illegal. Wood tanks are also fussy to build and have relatively short service lives. Anyone considering brick would do well to consider concrete block as a simpler, cheaper alternative.

Choosing materials

For the average do-it-yourselfer, the choice of materials practical for the construction of a septic tank is limited: wood, concrete block, brick, or poured concrete. Determining which is the least costly for a given size and design of tank is difficult, as is assessing which kind of tank requires the least amount of labor and skill. Much depends upon your experience and expertise with the various materials, the delivered cost of those materials, and your locale. But while you cannot easily estimate the labor involved, at least you can price the materials easily.

To calculate the quantity of concrete needed to pour a tank (FIG. 4-1), establish the inside and outside dimensions, then calculate the volume contained by the set forms, by multiplying the length times the height times the thickness of all six sections (ends, sides, top, bottom) and adding them together. Multiply this by 110 percent to include a waste allowance.

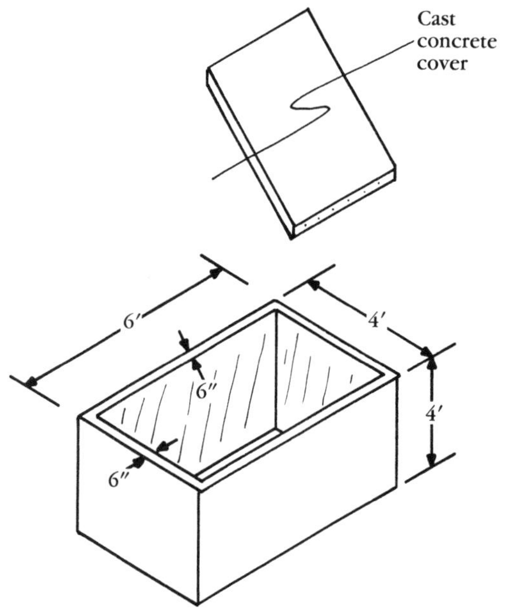

4-1 This concrete block or poured concrete tank with the dimensions shown has an approximate total capacity of 52.5 cubic feet, or 392.7 gallons. This means the tank has a working capacity of around 300 gallons.

If you use a single-sided form and depend upon the squared and leveled bottom and sides of the excavation for the other side, in your calculations only the top and bottom should be treated as though the form were double-sided. For the tank walls, determine the average distance between the position of the interior of the tank walls and the excavation sides. If the excavation sides are vertical and smooth, or nearly so, find the approximate volume and add 50 percent. If they are slanted, double the volume. Have something else (like a porch step) ready to pour in case you err on the plus side. To complete the calculations, add the cost of form lumber or rental unless you have a quantity of old boards or thick plywood around or can borrow forms.

To find the cost of a concrete block tank, first figure the cost of pouring a tank floor. Then find the total exterior wall area of the proposed tank in square feet. Multiply this number by 144 (the number of square inches in a square foot). Standard 8-by-8-by-16-inch blocks are fine for this purpose, so divide the result by 128 (the number of square inches of a block face). The answer is the number of blocks required for the walls.

You will need 4-inch-thick blocks to build divider walls if you select a multi-compartment design. You will also need sufficient concrete to pour the cover, steel bars for reinforcing both the concrete and the block walls if required, a quantity of mortar, and form materials for the floor and cover. You may prefer to buy a precast cover.

Excavating

You must decide whether to excavate with a pick and shovel or hire a backhoe to do the job. Even a small hoe can do the job in half a day. Even if you are in good condition and used to that kind of work, digging the tank hole and sewer line trench could take you several days.

Either way, a substantial amount of dirt will be displaced by the tank. Some of it can be left to one side for backfill. If the cost of hauling the excess off doesn't appeal to you, simply spread the spoil dirt over your grounds or use it for filling or recontouring. Wait to do this until you see how much will be used for backfilling the tank hole and how much the fill settles over time.

The depth of the excavation will depend upon the height of the tank you plan to build, plus gravel, plus 1 foot of soil cover, or as much as 2 feet in cold areas. The angle of the walls will depend upon the depth of the hole and the nature of the soil. In clay you can dig straight down and it will remain in place (if it doesn't get water-soaked). In soft or sandy soil you must angle the walls, perhaps even shore them up, to prevent them from caving in on you and your work. That can (frequently does) occur without warning, so play it safe to the point of overcaution.

If you are going to pour a concrete tank and the soil is unstable or soft you'll need to construct a double-walled form. The form walls are backed by studs and wales or walers (FIG. 4-2), so the outermost portions

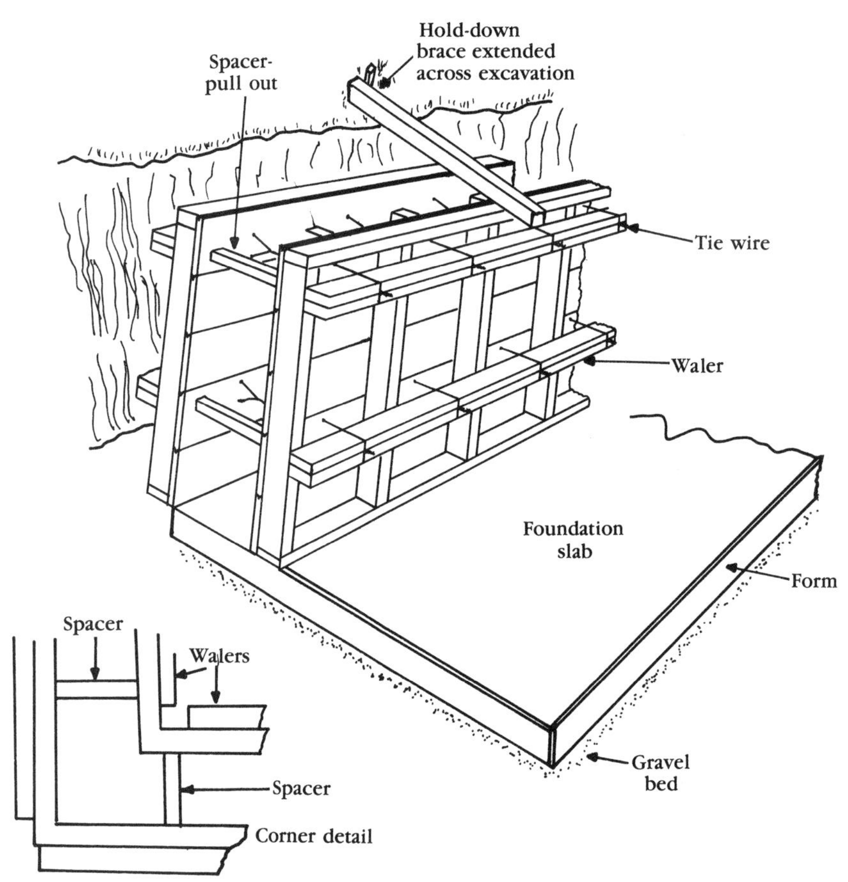

4-2 Setting up the forms for a poured concrete tank. The bottom, supporting slab, has already been cast. Corner framing is shown in plan view, lower left.

of the form will project 10 inches or so beyond the walls to be poured. Working room will also be needed for assembling the forms, so the hole for a poured concrete tank should be at least 3 feet larger all around than the tank dimensions (somewhat less is needed when commercial panelized forms are used).

If the soil is very firm and the tank height reasonably low, it is sometimes possible to use a single-walled form for a poured concrete tank (FIG. 4-3). The concrete is poured between the form wall and the dressed and dampened excavation walls. As much as double the amount of concrete may be used, but overall construction is faster and may be cheaper. Only half as much form and stripping materials is required.

BUILDING A CONCRETE BLOCK TANK

Excavate the required hole and cover most of the bottom with a layer of ¾-inch washed gravel. The gravel should extend a few inches beyond

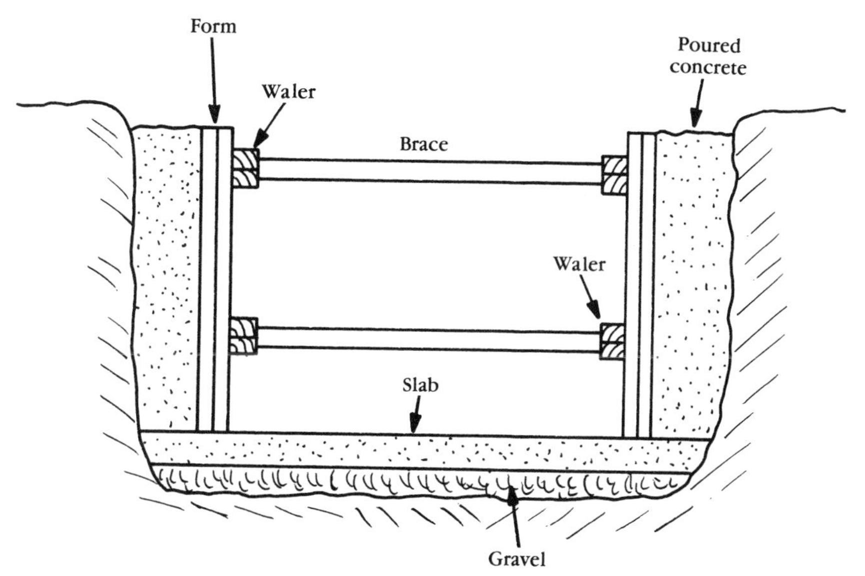

4-3 When the soil is firm, a single-wall form can be erected and the concrete poured slowly between it and the walls of the excavation. The supporting slab was poured before the wall forms were erected.

the planned exterior edges of the tank. Make the surface of the gravel as smooth and level as is practical.

The bottom of the tank rests upon the gravel bed, or cushion. Its dimensions are exactly the same as the desired exterior dimensions of the tank. If you used TABLE 3-1 as a guide for sizing, bear in mind that the dimensions given are for the interior of the tank. You have to allow for the wall thickness to get the overall dimensions.

Next, build a wood form for the poured concrete base slab. If the tank will be no more than 4 feet wide and 6 feet long, you can make the slab 3½ inches thick and use 2 by 4s on edge for the form. For a larger tank, go to 2 by 6s for the form stock, making the slab 5½ inches thick. In either case, reinforce the slab with a layer of 4-by-4- or 6-by-6-inch welded wire mesh (called re-mesh). Wear gloves and use a bolt-cutter to trim the mesh; cut it 4 inches or so shy of the slab edges.

Nail the 2 by stock into a rectangle, as shown in FIG. 4-4. Drive several stakes into the ground along the outside of the form planks to hold them in place. Make sure the form is level. The corners must also be square (perfect right angles). You can check this by measuring the diagonals. When they are equal, the form is in square.

Pouring the slab

Calculate the volume of concrete you will need by multiplying the length times the width times the depth of the form. Add 15 percent; a fair amount of concrete will settle into the gravel cushion. For an average

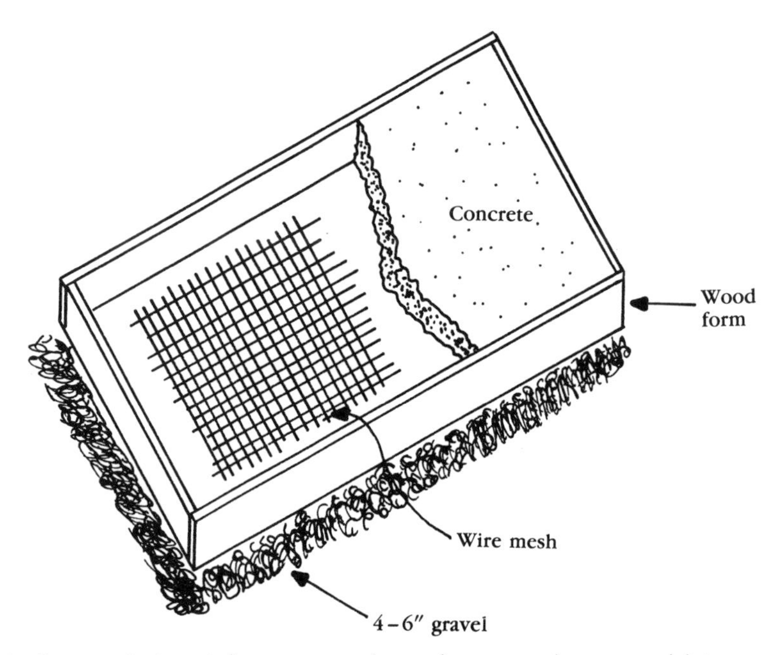

4-4 Form and wire reinforcement used to make a poured concrete slab to support a concrete block or brick tank.

sized tank, the amount needed is typically less than 1 cubic yard. Although ordering from a ready-mix company is certainly the easiest way to proceed, small quantities of concrete are usually quite expensive because of a minimum-charge levy.

You can also mix by hand using a good-sized mortar pan, or use a power batch mixer. Either way, use a 1:2¼:3 mix consisting of 1 part portland cement, 2¼ parts sand, and 3 parts aggregate (¾-inch gravel is fine).

To make 1 cubic yard or 27 cubic feet of this mix requires 6 cubic feet of cement, 14 cubic feet of sand, and 18 cubic feet of aggregate, plus about 36 gallons of water. The latter varies somewhat depending upon the dampness of the sand.

In case you are wondering how almost 43 cubic feet of materials shrinks to 27 when mixed, the sand fits into the voids between the stones, the fine concrete powder fills the voids between the sand granules, and the water fills all that's left. A standard bag of cement contains just 1 cubic foot. Use a graduated pail or similar container to measure quantities in correct proportions for a strong, workable mix.

You will have to mix the concrete in small batches, according to the capacity of your tub or mixer, but if possible the entire slab should be poured in continuous batches, not separated by several hours or days. To hand mix, put the cement into a tub or wheelbarrow first, sand on top, and dry mix them. Put the aggregate on top of that, and dry mix again until the mix is entirely uniform in color and consistency. Make a hole in the center of the pile and add water slowly, mixing meanwhile, until you have added the correct amount of water. Mix until all the ingredients are completely wet and the mix is plastic, no clumps or gobs.

To power-mix, start the machine and run in about 10 percent of the required water. Then add the ingredients in rotation, a shovelful of cement, one of sand, one of aggregate, a gallon of water, go around again, and so on. Move right along, and when the mixing has continued for about 3 minutes after you've finished adding materials, the batch is done.

In either case you will get stronger concrete if you use only the minimum amount of water to achieve a plastic, pourable, workable mix than if you pour soup so runny it can find its own level. A thin mix is weak and hard to work. Add a water-resistant agent, or some additive meant to increase water resistance, to make a more watertight concrete. Make your pour during good weather—45 to 85 degrees, moderate to high humidity, calm winds, as little direct sun as possible.

If you will use re-mesh, set it in the form. Dump the concrete into the form, starting at the far end and working back toward the near end, and also back and forth across the narrow dimension of the slab. Try to ensure that no more than about half an hour passes before a fresh pour is made against the edge of any previous one to avoid poor bonding and a weak point that could crack open later. Settle and level each batch after you pour it, and pull the re-mesh up to about the midpoint of the slab thickness. If you must pour the slab in separate sections, set form boards to divide the slab into smaller areas, stair-step the edges, and insert steel rod keys to tie the sections together (FIG. 4-5).

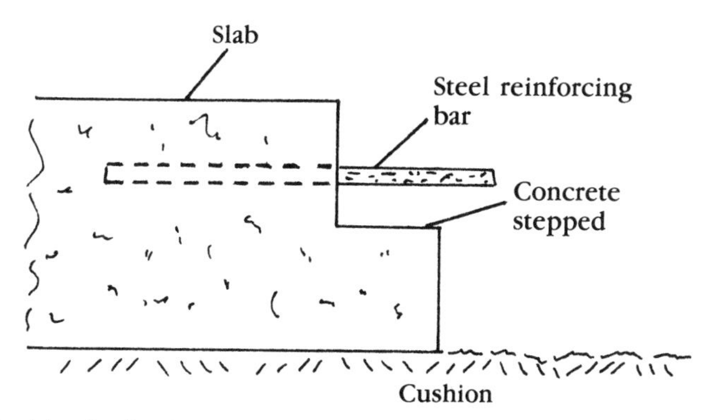

4-5 This slab edge has been stair-stepped and has steel rod keys inserted to solidly tie the edge to the next section of slab.

Once the form, or a smaller form section, has been filled, use a screed on the surface of the concrete to bring it flush with the top edges of the form boards; this is done by drawing a screed, or straight length of 2 by 4, across it with a sawing motion. Scrape the excess over the sides of the form. If there are low spots, scoop some of the extra back in and relevel those areas.

Then darby the surface. This involves gently floating a darby, which you can make yourself from scrap lumber (the wide flat surface of a 2 by 4, with a long handle attached, will do), across the entire surface. Use a slightly sawing motion with no downward pressure, moving the darby in sweeps. This knocks down protruding pebbles and smooths the small ridges left by the screed, consolidates the surface, and brings a bit of free moisture to the top.

The next step takes place after about 1 to 2 hours, when the sheen of surface water has just disappeared and the concrete is starting to cure in earnest. Equip yourself with a couple of 2-foot squares of plywood, one to kneel on and the other to put under your toes. With a wood float, which can be a 12-inch chunk of 2 by 4 or something fancier, start at one end and work backward across the slab. Sweep the float back and forth and around and about, smoothing and packing the concrete surface with just a bit of pressure; keep the float tilted a bit so it won't dig in. You should bring up little if any moisture.

Finally, cover the entire slab with 6-mil plastic sheeting. Secure the edges of the plastic with boards or rocks so it stays sealed and can't blow off. Leave it for at least 3 days, 5 if you can. By that time the concrete will have the greater part of its strength. Then remove the plastic and strip the forms away.

Laying the block

Use standard 6- or 8-inch concrete blocks, not cinder or other lightweight blocks. You can use all stretcher blocks, which have flanged ends, or switch to corner blocks (squared ends) at all the corners. This doesn't matter, because nothing will be visible.

Prepare the mortar by mixing mortar cement (which contains about 25 percent lime) with sand in the ratio of 1 part cement to 2 or 3 parts sand. Or use bagged mortar mix. Mix dry and then add sufficient water to produce an oatmeal consistency. Mix no more than you can use in about an hour, less in hot, dry conditions. If you do not use all the mortar in this time, discard it. If the mix begins to stiffen during this period, mist a *slight* amount of water over it and remix briefly.

With a mason's trowel, spread a bed of mortar along a slab corner edge about 1 inch wider than the block width, about 1 inch thick, and three or four blocks long. Make a furrow down the center of the mortar with the trowel tip. Align the first corner block and lower it gently onto the mortar bed. Push it down until the mortar joint thickness under it is about ⅜ inch thick (standard blocks are meant to be laid with ⅜-inch joints). Level the block in both directions with a spirit level, and make sure it is plumb (sides perfectly vertical). Make adjustments by gently tapping the top or sides of the block with the end of your trowel handle. Do not make any major adjustments that might break the mortar bond loose at any point. If that happens, take the block up, scrape off the mortar and lay fresh, and start over.

Stand the next two or three blocks on end and butter just the ends of the flanges with a layer of mortar about ¾ inch high. Set the second block in place by lowering it down and in toward the first block, mortared flanges first. It should come to rest aligned with the first block so as to create a ⅜-inch bed joint and a ⅜-inch flange joint, simultaneously. Again, level and plumb the block, check alignment, make adjustments with the trowel handle, and immediately set the next block or two in the same way. Scrape the excess mortar off the joints with the trowel, inside and out.

Next set two or three stretcher blocks extending the other way from the first corner block. Then spread a shell bed of mortar about 1 inch thick on half the first corner block and half of the opposing stretcher block. Set the second-course corner block into the mortar, and plumb, level, and align it. Then add more stretcher blocks in both directions, making a corner pyramid with the last block a corner of the third or fourth course (FIG. 4-6). Go on to the other three corners and repeat the process.

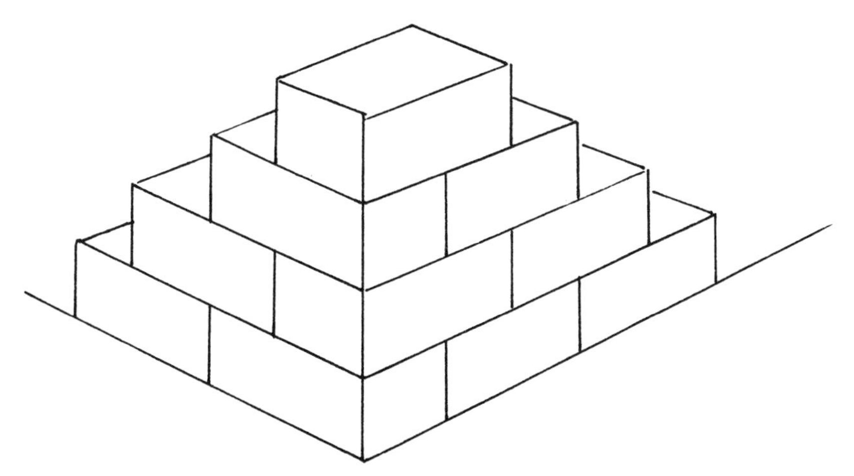

4-6 Corner pyramids of concrete blocks set on a slab start the construction of a block tank.

Using a pair of bricks or chunks of block as weights, stretch a taut line from one corner block to another. Set the line just outside the front edges of the blocks. This is the guideline. Lay down a full mortar bed on the slab from block to block. Butter the ends of a number of stretcher blocks and lay them aligned with the taut line. Scrape the joints flush as you lay the blocks.

If there isn't space enough for a full *closure* block to complete the course you will have to cut one. Lay a block flat on its side on a flat, solid surface (such as the slab) and score a cut line with a mason's chisel on both sides of the block. Then tap along the lines with a hammer to make the trim cut. Butter its edges, as well as those of both adjacent blocks, with mortar. Lower the block straight down, aligning as you do so, until the block beds (FIG. 4-7). If the closure block edges are ragged breaks you may have to fill the core with mortar to complete the closure. When

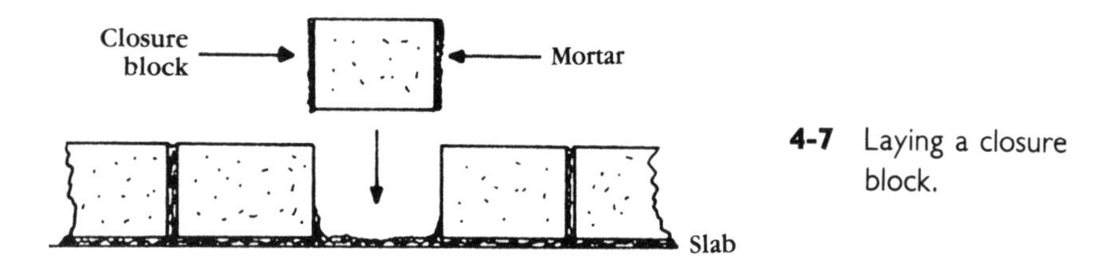

4-7 Laying a closure block.

doing this on upper courses, stuff the open cores of the course below with crumpled newspaper or scrap fiberglass insulation to prevent mortar from dropping into them.

Repeat this operation until you have filled out one tank wall—which one doesn't matter—and then go on to the others. Build up only four or five courses, then give the mortar joints time to cure. As you are working and the joints begin to cure, go back and tool them by pressing the damp mortar in fairly hard with a joint tool or a piece of ½-inch rod. This makes a concave joint that is dense, hard, and relatively moisture-resistant. To complete the tank walls, build up the corner pyramids and then fill in the courses in the same way. To accommodate the pipes, just leave gaps for them. Later you can position them and lock them in with mortar.

Use standard 4-inch-thick concrete block, sometimes called cap block, stood on edge for divider walls. Lay the blocks up in the same manner as the outside walls, using the slab as the foundation. Because there will be very little pressure on the dividers, you need not make them integral with the tank walls. However, tying them in does no harm, and is simple enough to do: this makes a stronger construction. At every other course, mortar strips of hardware cloth or similar screen right into the main wall joints so they lap out for 6 inches or so into the divider wall joints.

Plastering

Coat the interior of the tank with cement plaster to increase its life and ensure good watertightness. You can use the same mortar mix that you laid the block with. However, it's a good idea to mix in a little additive to increase water resistance to the mix, and to "fatten" it with a little extra portland cement. This will speed setting, make a more watertight coating, and a more workable plaster.

Using a steel trowel, apply the plaster to the slab a few inches in from any wall. With a sweeping motion, spread the plaster up the side of the wall (FIG. 4-8). Try to maintain a thickness of about ⅜ inch. When this coat has begun to cure, roughen the surface with a coarse, stiff-bristled brush or a scratcher made up of a handful of nails driven through a short length of board. Let the scratch coat of plaster cure for three days or more, then apply a second ⅜-inch layer.

The floor of the tank should be given the same treatment. Stepping on fresh plaster will ruin it, but you can work by sections, standing on a square of plywood on partly cured plaster as necessary.

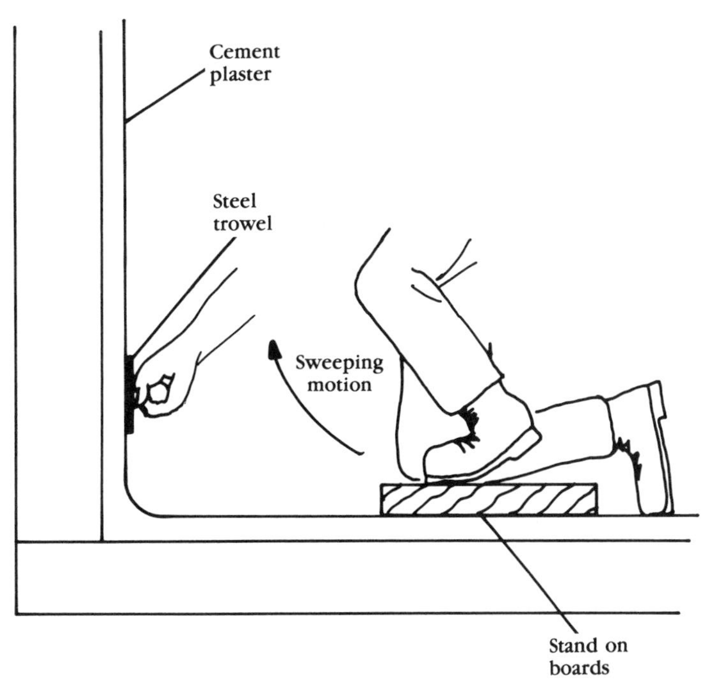

4-8 The inside of the block or brick tank should be cement plastered to extend its useful life.

Making the cover or top

You can make the cover of reinforced precast slabs (available at concrete product supply yards), you can use 3-inch bluestone slabs, or you can pour the cover of concrete, either in place or in a form nearby and then slide it on. It doesn't have to be a single slab of concrete; you can make several. The choice depends upon tank size, availability of precast products or other materials, and cost.

Before you position the cover, no matter what materials you use you have to make certain that the top edges of all the tank sides are flat, level, and equal in height. You can do this with a spirit level attached to a long, straight board. If you find low spots, build them up with mortar. To keep the mortar from dropping down into the block cores, stuff the cores with crumpled paper. The entire top surface of the tank walls must be in the same plane because none of the materials you might use for a cover will bend to conform to any irregularities. You must have a relatively air- and watertight seal here.

Precast slabs

When they are of the correct size and assembled, precast slabs do not have to cover the tank exactly, but can overlap without problem. Plain precast slabs present only the difficulty of cutting in an access hatch.

However, you can have such a slab made for you, and you presently can find a suitable ready-cast product at a septic tank supply yard.

In any case, spread a layer of mortar on the top edge of the tank walls to secure and seal the cover slabs. Then add a stripe of mortar atop the joints between slabs to seal the cracks.

Casting your own slabs Construct one of more forms from 2 by 4s nailed on edge to a sheet of plywood. Make the inner dimensions of the form such that an even number of slabs will cover the tank. Coat the inside of the form with form release oil so you will have no difficulty separating the form from the slab.

Use a 1:2¼:3 concrete mix that is just a bit wet. Mix thoroughly and cover the entire bottom of the form with about 1¾ inches of mix. Then lay reinforcing rods, selected according to TABLE 4-1, in place on the concrete and top up the form (FIG. 4-9). Screed the concrete to level it with the form edges and then darby it. Cover the slab with plastic sheeting, securely anchored, and allow 7 days for curing. The concrete will have about 85 percent of its strength by then, and you can safely place it on the tank.

4-1. Reinforcement Used with Cover Slabs

*Span**	*Bar size*	*Bar spacing*	*Crossbar spacing*	*Crossbar size*
4′	⅜″	10″	18″	⅜″
5′	⅜″	8″	18″	⅜″
6′	⅜″	6″	18″	⅜″
8′	½″	7″	12″	⅜″

*Slab to be a nominal 4 inches thick.

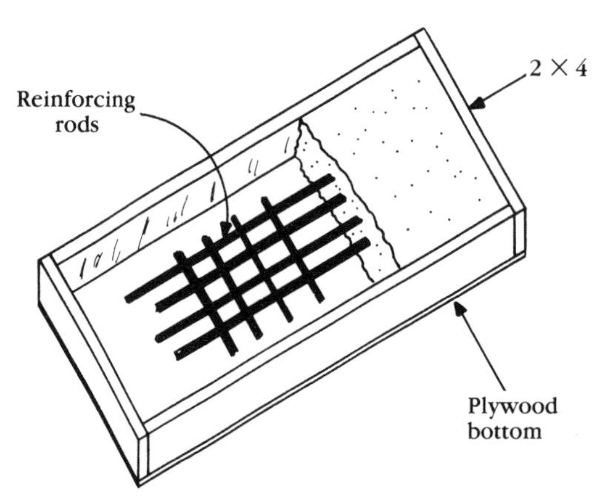

4-9 How 2 × 4s can be used to pour a tank-top cover. Reinforcing rods are required.

Providing an access hatch When you cast your own cover slabs you can incorporate a short section of large diameter vitrified clay pipe in one or more of the slabs (FIG. 4-10). Concrete pipe set the same way provides a better solution, because it is less susceptible to damage and you can purchase (or cast) a matching lid. Use pipe with a minimum inside diameter of 12 inches; local regulations may require a larger size. The length of the pipe can be calculated so the cover lies at or just below grade, but a preferable arrangement (sometimes required) is to extend the pipe about 12 inches or more above the finish grade. An alternative to casting-in a pipe section is to cast an opening in the slab and mortar a section of pipe to it afterward. Some locales require two access hatches, above the outlet and inlet pipes. Individual compartments also require separate hatches.

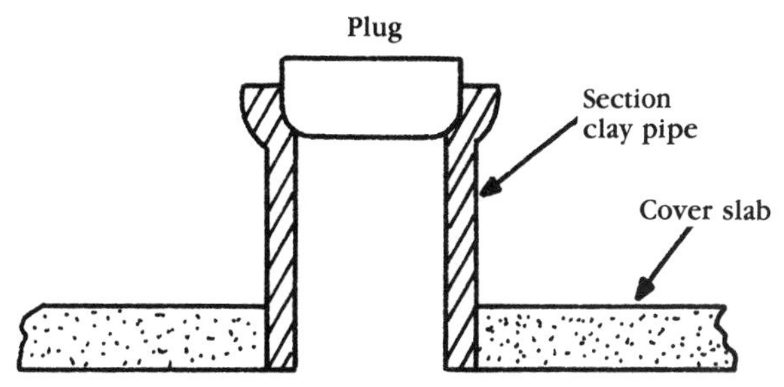

4-10 Making a port in a tank cover: The clay (or concrete) pipe section must be large enough to accommodate a long-handled shovel, or in some cases, a person.

Monoslab cover There is still another way to provide a cover for a masonry septic tank. You can pour it in place in a single operation. This is the best method, because it produces a solid, cemented-in-place slab that helps strengthen the tank. It also eliminates any off-level, out-of-true, or uneven wall top problems.

The techniques consists of constructing a wood false floor inside the tank, then surrounding the tank with a frame of wood. The concrete is poured atop the floor and is held by the frame form (FIG. 4-11). For slab thickness and necessary reinforcing rods, refer to TABLE 4-1.

To support the false floor, build a frame of 2 by 4s that fits inside the tank. Support the frame on more 2 by 4s spaced on 12-inch centers. Install crosspieces every 12 inches, running horizontally from one side to the other across the narrow dimension of the tank. Support these crosspieces with additional vertical 2 by 4s spaced about every 2 feet. Bear in mind that you will be holding up several thousand pounds of weight. Cover the crosspieces with 1 by 6 boards or sheets of, a minimum, ½-inch plywood sheets.

To provide access holes, make box frames and secure them to the form floor at appropriate locations. At least one hole must be comfortably large enough for a person to climb in and out of.

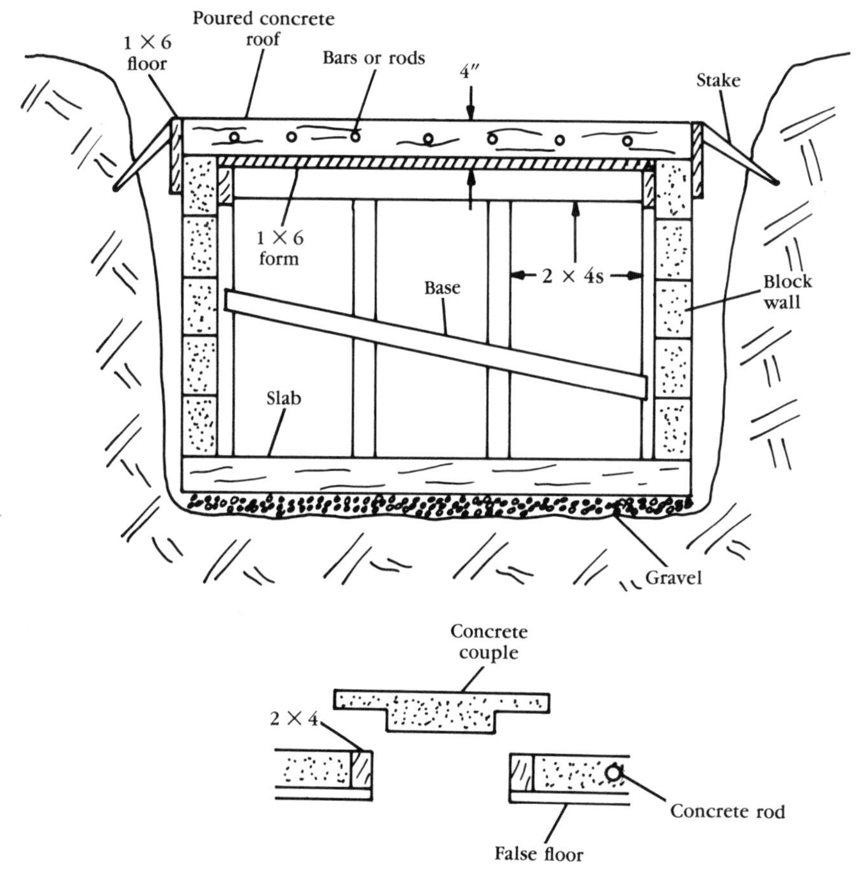

4-11 A. Pouring a slab atop a block tank. The wood form is removed after the slab has cured. Access is through the porthole. B. An aperture can be left in the poured slab to accommodate a concrete porthole.

Coat the interior of the form with form release oil. Pour a couple of inches of concrete over the entire form floor and rough-level it. Then set reinforcing rods in place and pour the balance of the concrete. Screed and darby the surface, cover with plastic sheeting, and allow 7 days for the concrete to cure. Then you can safely enter the tank and remove the form materials.

THE POURED CONCRETE TANK

The poured concrete tank is the best of the bunch. Made properly, it will probably last longer than any other type. The construction process starts with pouring a floor slab, just as previously described for the concrete block tank. There are, however, a couple of minor modifications to consider.

First, the walls should be poured within a few days of pouring the floor slab. If that isn't possible, coat the area where the walls will join the floor with a bonding agent. Otherwise, bonding of the joint between the fresh and the cured concrete may be weak.

That gives rise to the second modification. As you level the fresh concrete of the slab, press a continuous series of 2 by 2s, end to end, into the concrete all the way around the slab about 3 to 4 inches in from the perimeter (depending upon the thickness of the walls to be poured). Their surfaces should be flush with the concrete surface, and they should be treated with form release oil. When the concrete has cured, pull the 2 by 2s up. This will leave a keyway that will lock the concrete walls to the floor.

Building the wall forms

If the walls of the excavation are hard and firm soil, relatively vertical, and the hole is not too deep, you may be able to build a single-walled form and pour the concrete between the soil and the form (see FIG. 4-3). But if the soil is too soft and porous or the hole is deep, as is usually the case, you will have to build a conventional double-walled form like that in FIG. 4-1 (or erect a similar commercial form system). Use 2 by 4s for the studs and back them with horizontal 2 by 4 members called walers. Their number and spacing depends upon the height and thickness of the form and the thickness of the form boards. Height and thickness, along with the rate of pouring and the temperature, determine the pressure developed by the fresh concrete at the bottom of the form, and it can be enormous. At a pour rate of 5 feet per hour, for example, at 50°F, the concrete exerts a pressure of over 1000 pounds per square foot. At that pressure both the stud and wale spacing should be about 12 inches on centers. Your forms must be tight, very strong, and stoutly anchored and braced.

Because concrete begins to cure fairly rapidly, you can reduce this pressure by pouring at a slower rate, going around and around the wall forms and building the level up gradually. Also, the warmer the ambient temperature the less pressure generated. You should not, however, stop pouring until the form is filled, the concrete rodded or vibrated, and the top screeded level. The pour should be one homogenous mass.

The form walls should be made of ¾-inch plywood or special concrete form plywood if the walls are 5 feet high or more. Below that you can use ⅝-inch or even ½-inch plywood, or 1 by 6 tongue-and-groove boards. The interior surfaces should be treated with form release oil. Place tie wires and spreaders between the form sides, as FIG. 4-12 shows, to maintain alignment. An alternative is to install any of several kinds of combination tie-spreaders that are commercially available. There are differences; follow the maker's instructions for installation.

Calculate the amount of concrete required on the basis of the form volume (height × thickness × length) plus 10 percent for waste. If you require more than a couple of yards, you probably will be well advised to order ready-mixed concrete. This makes the job much faster and easier, likely will be less expensive, and you'll have a far more uniform

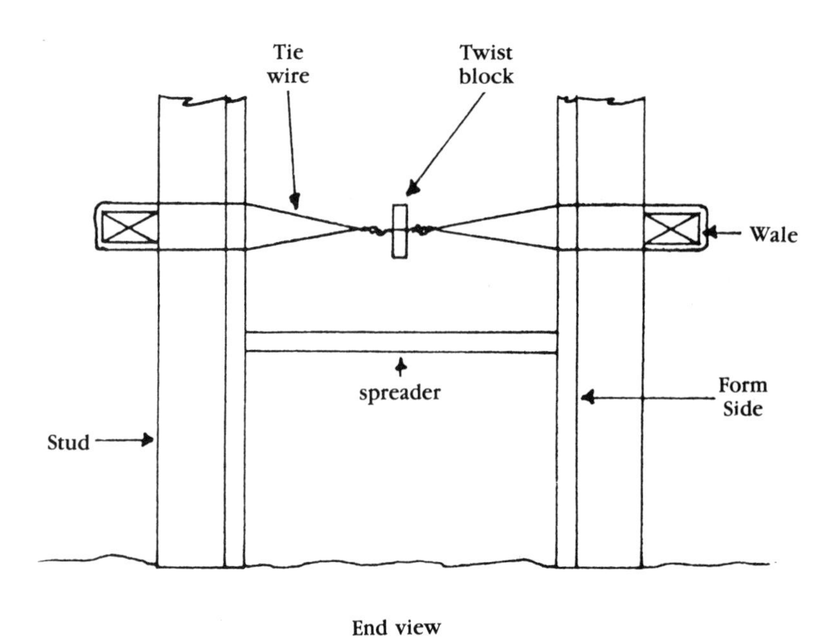

4-12 Wire ties and wood stick spreaders will maintain the alignment of the form walls. Pull the wood spreaders out as the concrete is poured.

pour of better quality. Tell the company the purpose of the concrete and request a waterproofing additive, and they'll send you the correct mix.

It's best to have a helper on hand when the truck arrives so the job moves along smoothly. To make certain the concrete has reached all the corners and there are no air bubbles, scrape a flat-bladed shovel or an ice chopper up and down the inside of the form walls as the concrete is poured, and shove a length of 2 by 2 up and down in the mix, like churning butter, to settle it and break up air pockets. Rapping the outside of the form with a hammer is another method. Professionals use special vibrators, but this requires some expertise as it can easily be overdone.

After the pour is complete, cover the wall tops with plastic sheeting weighted with boards or whatever. Give the concrete 3 or 4 days to cure, then carefully strip the forms. If you strip too soon you'll damage the green concrete, if you wait too long after pouring you'll have trouble peeling them off. Wait a few more days before backfilling, and do so carefully, with clean fill containing no rocks larger than fist-size, if possible. The cover can be poured or installed any time after about 7 days have passed.

The same two layers of mortar cement previously described for use with the concrete block tank can be applied to the inside of the poured concrete tank. Allow several days for the plaster to cure.

Chapter 5

Piping

Sewage is routed to the house main drain, a single pipeline that extends through the house foundation. Here it connects to the sewer line, which carries the sewage to the septic tank. Outflow from the septic tank runs through a single solid-walled pipe to, in most cases, a distribution box. From the box two or more leach lines of pipe extend into the leach field. Several kinds of pipe and the associated fittings might be used for these purposes; one particular kind may be required in some locales.

PIPE TYPES AND DIMENSIONS

While the building and plumbing codes throughout the country are far from being carbon copies of one another, they are fairly close in many details, such as the diameter of a house main drain. Codes usually call for a 4-inch pipe (meaning a pipe having a 4-inch inside diameter). But because copper and plastic, now almost universally used, have inside walls so much smoother and have so much better flow characteristics than the old conventional cast iron pipe, most codes now permit 3- or 3½-inch pipe to be used. Even so, it is a mistake to install anything smaller than 4-inch pipe as a sewer line, even if the house main drain is 3- or 3½-inch pipe. Installing the larger size reduces the possibility of blockage and freeze up in a crucial section of pipe that can be difficult to clean. A typical transition installation of house main drain to septic system is diagrammed in FIG. 5-1.

Choosing materials

Many still feel that the best material is cast iron. Although brittle (by comparison with steel), it is strong and its thick walls ensure longevity. This may indeed be the longest-lived pipe, though that has not been

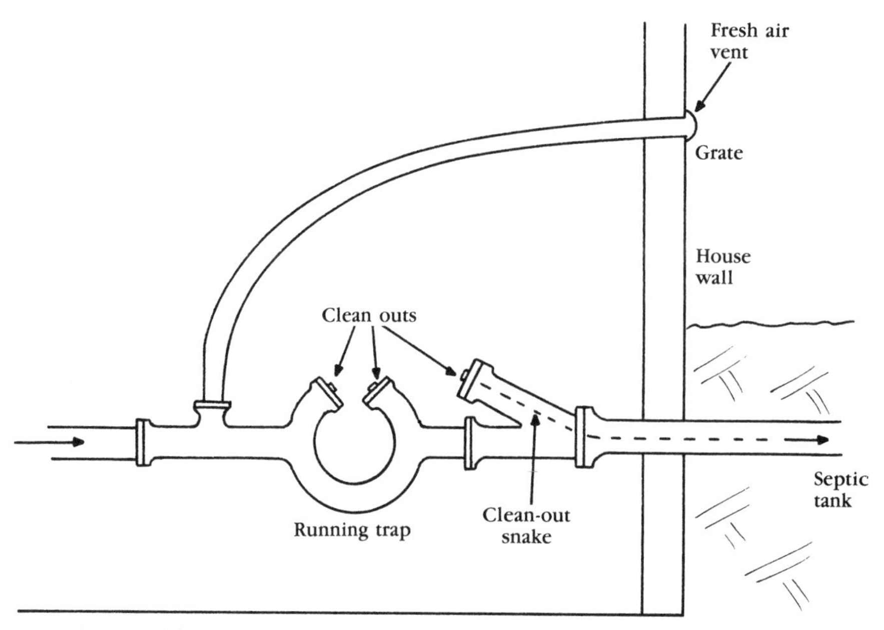

5-1 A typical house main drain to sewer line connection. Air vents and running traps are not always used.

proven. In some homes, cast iron sewer pipes and vent stacks have been in continuous use for more than a century, and in some locales it is still required. Cast iron pipe is very heavy, available in 5- and 10-foot lengths in three service weights, difficult to cut, and difficult to join. The joints must be leaded together, which requires special tools and expertise.

Always use cast iron pipe when the line must run under a concrete floor in a basement, at least in the under-concrete section. It is also a good idea to use it where the pipe passes under a walkway, driveway, or road.

Copper pipe is widely installed in residential drain-waste-vent (DWV) systems, though less nowadays than previously because of the high cost. But it should not be installed beneath concrete or buried in the earth, and it is not used in sewer or leach field lines.

Transite pipe is made from concrete and mineral fibers. It is fireproof, rotproof, bugproof, and will not corrode. It is nowhere near as strong as cast iron, but is lighter and easier to work with. It can shatter under a heavy load, but is otherwise very long-lived. It has been popular in some areas for septic tank work and serves adequately.

Vitrified clay pipe is a traditional type still is limited use. It can be laid tightly joined as a sewer line and loose-jointed as a leach line. It is available typically in 2-foot lengths. Clay pipe handles easily enough but is difficult to work with, fragile, and the joints are weak at best. They must be made with oakum and mortar. Barring structural damage, clay pipe is extremely long-lived.

There are two kinds of plastic pipe that are now in widespread use for both sewer and leach lines: PVC (polyvinyl chloride), and to a lesser extent ABS (acrilonitrile-butadiene-styrene, if you really must know). Typically available in 10-foot lengths, several diameters, rigid and both solid-walled and perforated, this pipe is lightweight and easy to handle, very easy to work with, tough, and long-lived. It is impervious to practically everything except severe mechanical damage and certain plastic solvents, and is relatively inexpensive.

Fittings

Fittings are defined by their dimensions. Thus, a 4-inch elbow or tee accepts a 4-inch pipe at each opening. However, the openings may be of different sizes, which are used in sequence to designate the fitting. Thus, a 4 by 3½ by 4 tee, or a 4 by 3 elbow. Fittings are also named by their shapes or purposes. A tee looks like a T, an elbow or el bends like an elbow, a reducer steps from one size to another, a coupling joins two lengths of pipe, a sweep el forms a sweeping, gradual bend rather than a sharp one like an elbow.

In most cases the dimension given refers to the inside diameter or trade size of the pipe. Sometimes it is the outside diameter, depending upon the pipe material. Also, the same kind of pipe may come in different wall thicknesses, which result in greater outside diameters, such as Schedule 40 (lightweight) and Schedule 80 (heavyweight). So it is important to purchase fittings that are designed for use with particular types of pipe. Otherwise the joint will be weak or leak, or the fitting will not mate up with the pipe at all.

Usually the fittings are also made of the same material as the matching pipe, but not invariably. Transition fittings are made to connect pipe made of one material to pipe made of another. Examples are cast iron to plastic or plastic to copper. An example of an entirely different fitting material is the rubber or neoprene sleeve coupling that can be used with plastic or cast iron pipe.

RUNNING SEWER PIPE

The term *running* in plumbing parlance means to cut, fit, and install pipe. The methods or techniques used vary with the kind of pipe. Cast iron and vitrified clay pipe are manufactured with *bell* and *spigot* ends. The bell is the enlarged end into which the narrower spigot of the preceding length of pipe fits. The other kinds of pipe discussed earlier are joined by various kinds of fittings that can collectively be called *slip* fittings.

Cutting cast iron

There are three ways to cut cast iron pipe. One is to score a line with a hacksaw—going around and around, deeper and deeper—until you have a groove about ¼-inch deep. Then a light tap with a hammer should cause the pipe to come apart on the line.

A second method consists of striking a break or cut line with a dull cold chisel and a hammer. Keep tapping lightly on the line, going around and around the pipe until it breaks on the line. To do this properly you have to support the pipe firmly on a block of wood or solid ground. The blade of the chisel compresses the fibers of the steel until they separate and the pipe itself comes apart.

The third method is the easiest, but it requires a pipe cracker. This is sort of a giant pair of pliers used to apply pressure to the pipe, breaking it along the preferred line—most of the time. You might be able to rent this tool locally, or perhaps borrow one from a local supply house or friendly plumber. The big advantage is that cutting with the cracker requires slight effort and little time.

Joining cast iron pipe

The usual method of joining cast iron pipe starts by inserting the spigot of one section into the bell of another. Both pipes must be perfectly aligned. With the aid of a caulking chisel (or a short length of suitably shaped hardwood), a layer of *oakum* is forced into the space between bell or hub of one pipe and the spigot of the other. Oakum is a soft, coarse rope made for this purpose. It is wound around the pipe before being forced into the opening. You need sufficient oakum to fill about ½ inch of the cavity, which seals the joint.

Next you fill the remaining space between the spigot and the hub with lead. You can use molten lead, melted in a lead pot and poured in with a ladle, or you can use *lead wool* (fine lead wire) and pound it in cold (see FIG. 5-2).

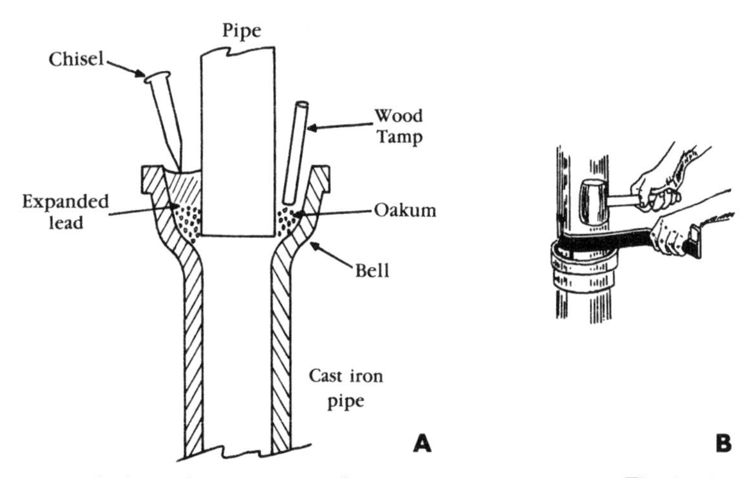

5-2 A and B show the steps in caulking a cast iron pipe joint. The lead can be poured (or in wool form, fine wire) and pounded in order to expand it. For a temporary joint or an iron-to-clay joint, mortar can be used instead of lead. If this is done, it is brought farther back along the spigot to make a deeper joint.

In either case, you need at least 1 inch of lead in the joint but not so much you fill the opening. Next the lead must be "spread." This consists of striking it with a dull chisel. The blows expand the lead and lock the two sections of pipe together. If the joint is not disturbed, it will remain intact and sealed for as long as the pipe lasts, and that's a long, long time.

An alternative method is to use hubless fittings. This requires removing the bell or spigot so you end up with sections of straight-walled pipe. The hubless fitting consists of a thick rubber sleeve and a pair of stainless steel worm-drive clamps. There may or may not be a stainless steel outer sleeve. The rubber sleeve is forced over the pipe ends and the clamps are snugged down to seal the connection and hold the pipes in place (FIG. 5-3). The advantage of the hubless system is that it is very fast (if you have a pipe cracker to trim the pipe ends); the joint is slightly flexible and will sustain light shocks; and the joint will bend a few degrees, something that is impossible with a lead or cement joint.

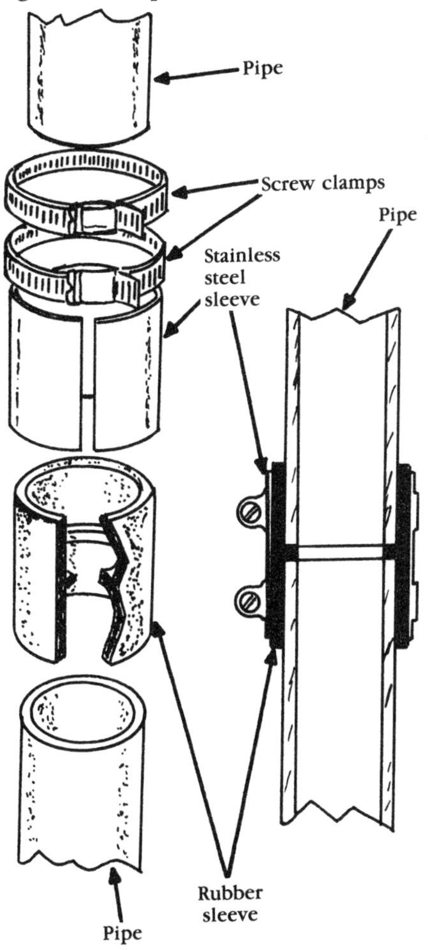

5-3 An exploded view of a No-hub joint. You can either purchase a length of hubless cast iron pipe or cut the hub off.

Still another way of joining cast iron pipe consists of using oakum followed by Portland cement mortar (FIG. 5-4). The spigot is positioned within the hub and the oakum is pounded home. This is followed by a layer of mortar made by mixing 1 part Portland cement with 2 parts clean, sharp sand and enough water to make a thick mix. This type of joint is deemed the least desirable of all the cast iron pipe joints. Its use is generally confined to pipe buried in the earth.

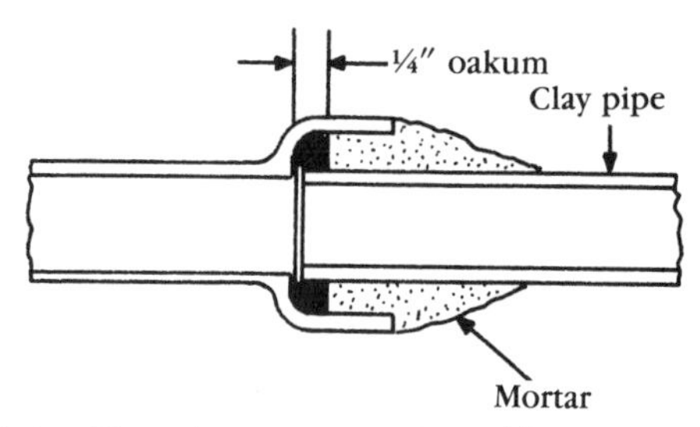

5-4 A clay pipe joint. The oakum must be about ½ inch thick and plenty of mortar cement must follow. This is a bell and spigot joint.

To run cast iron pipe a specific distance, that distance must either be an exact multiple of pipe length or you have to cut some pipe to make it fit. The problem that might come up is that you could end up needing a very short section of pipe to complete the run. Ideally, you would start the septic tank run with a spigot end inserted into the tank inlet, then add pipe sections until you reach the house, continue the run as the house main drain, and work your way right up the main soil stack and through the roof. The pipe sections simply follow one another from one fixed point (tank) to a variable one (above the roof). But this is generally impractical. Most often the house plumbing is completed first, and the pipe is later run from the house to the septic tank.

The solution is to run the first part of the line, but lay out the last several lengths dry. Then you can see just how much pipe you will need. If necessary, you can cut a little from several lengths of pipe and thus not end up requiring an impossible 3- or 4-inch piece.

Another problem that can occur with cast iron pipe is that you have to end up with a hub, or bell, facing the tank so you can use a standard fitting inside the tank. This can be accomplished by first running the sewer line all the way to the tank. Then the last fitting on the line, which will be inside the tank, is slipped through the inlet hole in the tank wall, spigot end first, and into the bell on the last section of pipe.

Vitrified clay pipe

Lay the pipe on a board or solid earth so the hub is in the clear and unsupported. With a cold chisel and hammer, score a break line all the way around the pipe. Repeat this again and again, turning the pipe meanwhile. Eventually the pipe will break apart along the line. The trick to the technique is to use light taps, not hard blows, on the chisel, and have lots of patience.

Make up the joints with oakum, followed by mortar, as explained for cast iron pipe. These joints are weak. The joined pipe sections can be positioned vertically without many supports, but if the pipe is laid horizontally it must be supported. Once positioned and joined, the sections should not be moved. Beneath the ground, the earth provides sufficient support. Above ground, as in a cellar, install concrete blocks or similar supports at each joint.

All the problems of pipe size and fittings that are common to cast iron pipe are encountered with clay pipe as well. Note that clay pipe can be connected to cast iron pipe with no danger of reaction or deterioration. When the spigot of a cast iron section is joined in a clay bell, the joint is made with oakum and mortar. When clay pipe is joined in the bell of a cast iron section, the joint can be made with mortar or hot lead, but the lead must be expanded very gently. Lead wool cannot be pounded in place.

Plastic pipe

You can cut large diameter plastic pipe with any ordinary wood- or metal-cutting saw. A fine-toothed crosscut handsaw works about the best. Wrap a straight edged strip of paper or a length of tape around the pipe to give you a straight guideline to follow with the saw. Remove burrs made by the saw teeth by scraping the edge with a file or knife.

Because plastic pipe can be cut to any length, there are none of the dimensional problems that can be encountered with clay and cast iron pipe.

Joining one section of plastic pipe to another, or to a fitting, takes less time to do than to describe.

To join two pipe ends, use a coupling of the same size and a compatible material. Prepare the surfaces by cleaning them and removing the shine with very fine sandpaper or steel wool, or better yet, clean them with the liquid pipe cleaner made for the purpose. Clean both pipe ends for a distance about equal to its diameter, as well as the inside walls of the coupling. You can also treat the surfaces with a primer, but this usually isn't worth the trouble.

Let's assume you are joining two pipe sections by means of a coupling or a slip fitting. A coupling has an inner lip that prevents the fitting from sliding entirely onto one section of pipe. A slip fitting has no internal lip, and can be positioned anywhere on the pipe. In any case, the rotation of one pipe section relative to the other is unimportant.

Be sure to select the proper pipe cement for the particular plastic your pipe is made of. Using the dauber attached to the can top, wipe a coating of cement on one pipe end and another on half the inside wall of the coupling. Work rapidly, and slip the coupling and pipe together. Make sure the pipe end seats against the internal lip of the coupling. At the same time, twist the coupling an eighth to a quarter turn. Let the cement set up for a couple of minutes, then repeat the process, slipping the end of the second pipe into the other side of the coupling.

Now assume that you are going to fasten an elbow or some other directional fitting on the pipe end. Here the rotation of the fitting in relation to the pipeline is important. Assemble the joint dry, exactly oriented. Draw or scribe a line on the pipe that continues onto the fitting. Disassemble the joint and apply the cement, then reassemble it. Give the fitting a fraction of a turn as you do so, ending up with the two marks aligned with one another, thus maintaining the correct orientation.

When a bead of cement appears all around the edges of a fitting, you have used the right amount of cement. More is just a waste, and in smaller pipes can build up ridges or lumps that impede liquid flow. The cement should be at room temperature (warm it in a pan of water if necessary), and should not be applied in cold weather or to cold pipes. If the cement has become "ropy" or thick, thin it with cement solvent, or even better, discard it.

The great drawback of plastic pipe is that once a joint has been solvent-welded, that's it—you can't take it apart. If a joint leaks or you've made a mistake, all you can do is cut the section out and replace it with a short piece and two slip fittings. You can, however, use hubless fittings to join pipes, and they are both adjustable to a certain degree and removable.

Transite

You can cut Transite with a hacksaw fitted with medium- or fine-toothed metal cutting blades. It is normally available in 5- and 10-foot lengths.

The pipe is manufactured with tapered ends designed to be forced-fitted into special couplings. These have internal neoprene O-rings that seal the joint. To install a fitting, align it carefully, place a block of wood square across the open end, and tap it home with a hammer. Make pipe to fitting joints the same way, keeping the pipe well aligned.

It is wise to avoid a situation that requires cutting Transite except as a stub end where no fitting is needed. This is because you must then attempt to file or grind the requisite fitting shoulder on the cut end (difficult and chancey), or find an external rubber sleeve or no-hub coupling that will fit.

Masonry yards and pipe suppliers rarely carry more than a few basic Transite materials, if any. If you require a considerable amount of pipe or fittings, especially if they are unusual, be sure to order them well in advance of the construction date.

Chapter 6

Planning the septic system

The positions to be occupied by the septic tank and the leach field are sometimes obvious and naturally suitable, and at other sites there may be only one choice. But more often, selecting their positions is a matter of compromise. It is important to know what factors must be considered and what weight should be given to each.

The best way to plan the layout of your septic system is to work with a combination plot and topographical map. Drawing one to rough scale is not difficult.

If you have a copy of the *plot plan*—the blueprint that shows the property boundaries and improvements, drawn up by a surveyor—you have all the major dimensions. Transcribe this plan onto graph paper, making it easier to keep the features in their proper relationships. If you do not have a plot plan, go to work with a steel tape and measure the grounds. Add the buildings and whatever else is present on the property, such as outbuildings, drive, well, etc., all in their proper locations (FIG. 6-1). Include the water service pipeline and the path of electrical and telephone wires if they are, or will be, underground. Verify that you don't run afoul of any easements, such as for telephone cables or high-pressure gas lines.

Adding the topography

The map in FIG. 6-2 is a rough plot plan with some contour lines added, indicating the relative elevation of the ground at various points. The road is the reference point at zero elevation (0′). The area immediately adjacent is fairly level and also at zero elevation.

As you leave the road, you see a line reading 1.0 feet. This is a contour line, and means that the indicated area all along this line is 1 foot higher than the road. Farther in the same direction, the next contour line is marked 2.0 feet. The ground along this line is 2 feet higher than

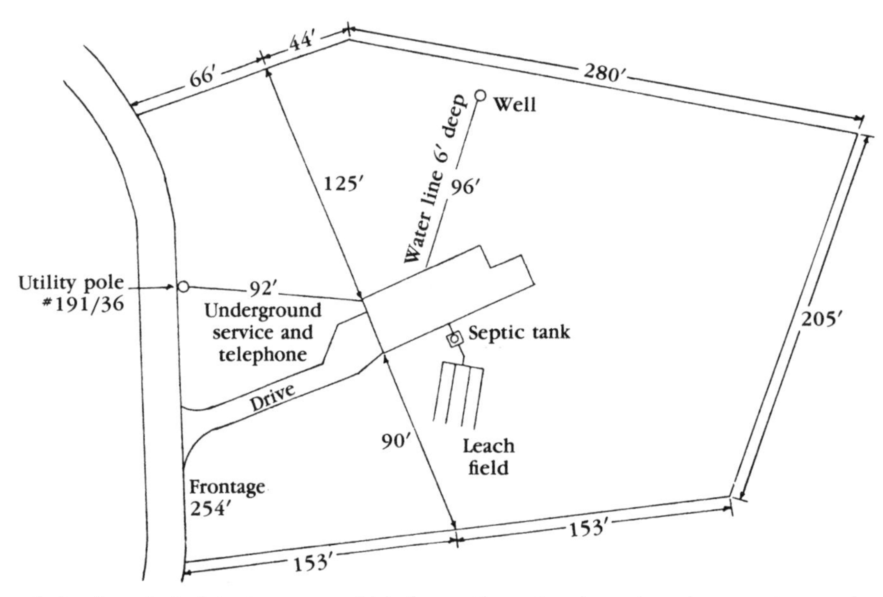

6-1 A typical plot plan, upon which the septic system layout can be superimposed.

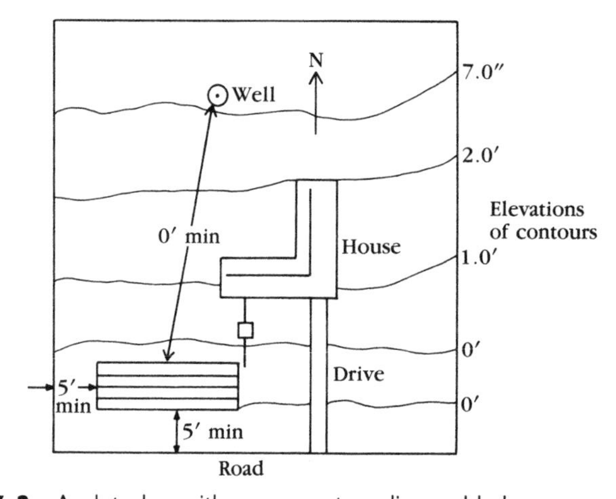

6-2 A plot plan with some contour lines added.

the road and 1 foot higher than that along the previous line, and so on. The closer together the lines are, the steeper the grade. Reading the contour line figures, you can see that the property ascends more quickly toward the rear, reaching a height of about 7 feet.

Mapping the topography of your building site is not difficult because it does not have to be especially accurate. Assuming that there is a pitch to the land, start with the highest point that is convenient. Drive a short stake into the ground. Attach a line to this stake, and hold the other end in your hand. Hang a line level from the line, move downhill, and stretch the line taut. Have a helper position the level more or less halfway between you and the stake. Move the line up or down until your helper tells you the bubble is centered and the line level. The distance from the string to the ground at any point along it is the elevation at that point.

Note the elevation on your map. The peg represents 0, string to earth elevation. Repeat at a number of points. On the map, draw lines connecting points of equal elevation (FIG. 6-3). Now reverse the elevations. Make the peg the highest point. The other points are lower.

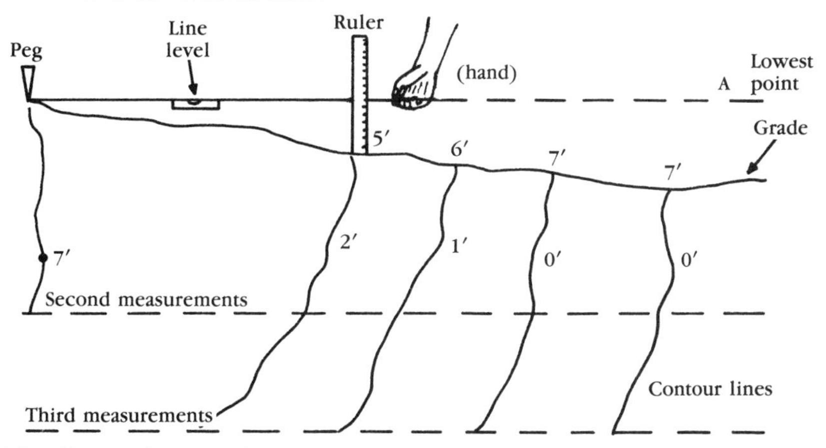

6-3 Contour lines are developed with a few simple instruments. Elevations over an area are ascertained and then transferred to a corresponding scale map.

With your map in hand, you can move your septic tank and leach field about until you find the best positions for them. The map makes visualizing the factors that must be accounted for far easier than doing the same while walking about on the property.

If the building department or local health authorities have a say in your tank and leach field, you might be required to present specific data on where you plan to place everything before they will approve the installation. Your map will show that information. If well-dimensioned and accurate, it can also serve as a layout and excavating plan.

THE SEPTIC TANK

The septic tank is the first element of the system to position. Typically it is placed about 10 feet—one length of plastic sewer pipe—from the house. The overall piping is usually laid out so that all pipe runs are as short as possible to save expense and reduce maintenance and the possibility of clogging. Ideally, the house main drain should exit the building in line with the most favorable location for the tank. If possible, the route from the end of the house main drain should be a short, straight run. If necessary, the sewer line can be angled, and the tank can also be placed much further away from the house (closer is not recommended). The tank alignment with the leach field should also be considered, and some compromise might be necessary for the best piping lineup. Make sure the tank is not in a setback area, or too close to a neighbor's fence.

Septic tank elevation

The elevation of a septic tank with respect to both the house and the leach field is crucial. At best it can be only a few inches off. The tank should be deep enough to permit 1 foot minimum of earth cover. In cold climates this should be increased to 2 feet or more.

The inlet pipe should be centered in a hole about 8 to 10 inches below the top edge of the tank. This is to provide sufficient room for a pipe tee and allow a clearance of an inch or two for venting. The inlet can be adjusted up or down a few inches, but no more. Lowering the pipe reduces the capacity of the tank, while raising it reduces or eliminates the venting space. The proper relationship of tank inlet to outlet must be maintained.

The pitch of the sewer pipe leading to the inlet is crucial. This line can be set at a downward pitch from the building to the tank of ¼ to ½ inch per running foot, no more and no less. Increase the pitch and the liquid will run off quickly, leaving the solids in the pipe. Lessen the pitch, and the solids will also remain in the pipe. If the pitch cannot be held to these tolerances, the line must be set to an angle of 45 to 90 degrees.

Bear in mind that the sewer pipe is not flexible. You can bend it only a fraction of a degree at the joints. If you try to cock the pipe any more, the joint will leak. You can, of course, use angle fittings, but they are limited to 22½, 45, and 90 degrees. The elevation of the septic tank must be such that you can make the sewer pipe run at either the required slope or incorporate one of the commercially available elbows.

Checking elevation

You can measure up from the top surface of the sewer pipe as it exits the building and transfer this dimension to the outside of the building. A more accurate method consists of digging a short trench at the outside of the building down the side of the foundation until you reach the sewer pipe. Then measure up from the top edge of the sewer pipe any convenient distance and mark the distance off with a nail driven into the wall.

Let's say you have measured up exactly 5 feet. Now go to the point where you want to excavate for the septic tank. Drive a stick vertically into the ground at the spot where you believe the inlet end of the tank should lie. Stretch a line from the nail in the house wall to the stick. Hang a line level on the line and adjust the string up or down against the stick until the bubble in the level centers.

You now know that you have to go straight down 5 feet from the string-stick junction to make the sewer pipe horizontal. But you need a pitch. To determine the correct amount, first measure the string from stick to house wall. This gives you the approximate required run (length) of the pipe. Multiply that figure by ¼ inch to find the necessary drop that will produce the desired pitch. Assume the run is 32 feet. Then:

$$\text{Pipe run in feet} \times \tfrac{1}{4}\ \text{inch} = \text{required drop}$$

$$32\ \text{feet} \times 0.25\ \text{inches} = 8\ \text{inches}$$

Bear in mind that the nail is 5 feet above the top of the sewer pipe. The tank end of the pipe has to be 5 feet plus 8 inches below the horizontal line represented by the string. To determine the required depth of the tank excavation, first measure down the stick from the string to the ground. That gives you the elevation of the surface of the ground (the grade) at that point in relation to the house sewer pipe. That in turn tells you how much soil there will be above the septic tank.

Knowing the vertical dimension of the tank you plan to construct or install and knowing where you want the sewer pipe to enter the tank, you can calculate the necessary elevation of the bottom of the tank based on the distance that will separate the tank from the house (the length of the sewer pipe).

When you actually excavate for the tank, you can double-check your figures with the aid of a vertical stick or pole positioned where the sewer pipe will enter the tank. Remember that you are working with the top surface of the pipe. To find the center of the inlet hole you must take half the outside diameter of the pipe plus the clearance allowance and add it to the measurement.

Tank rotation

The inlet tee fitting that you will install will come through the tank wall at a right angle. When at all possible, that tee should point directly at the house sewer pipe as it leaves the building, so there will be no bends in the line impeding the flow of sewage and providing hang-up points for the solids.

If the house sewer pipe absolutely cannot be directly aligned with the septic tank inlet, there is no recourse but to introduce bends in the line. The tank must be positioned at the correct distance and the inlet so oriented that either a single angle fitting will allow an otherwise straight run to the tank, or a pair of fittings can be installed as shown in FIG. 6-4.

One way to plan this is to draw the building, tank, and leach field to scale in block form on graph paper, showing their dimensional relationships. Using a protractor, you can then figure what fittings can be used with the tank and field placed at various positions relative to the house. Another method, sometimes easier and more foolproof, is to actually lay the pipe out on the ground surface. Use a length of board to represent the tank end, and shift the components around until you find a satisfactory alignment.

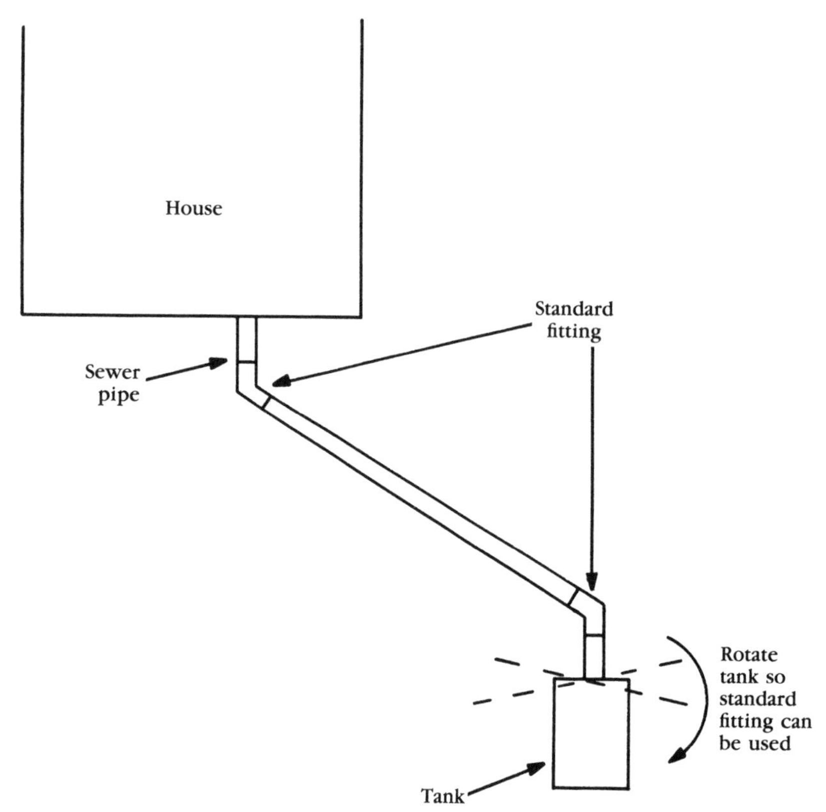

6-4 Where an offset sewer line must be installed, it is important to orient the septic tank so that standard angle pipe fittings can be used.

In some installations two tanks are required, one feeding the other. The purpose is to allow much more time for the raw sewage to be broken down and the solids to settle out. They can be placed side by side, or end to end, depending upon available space.

Sewer pipe length

Whether or not the distance from the end of the house main drain to the inlet end of the septic tank is crucial depends upon the material you plan to use for the sewer pipe. If plastic pipe will be laid, you can easily trim it to any length with a handsaw, and extensions are easily made with slip-on couplings. That is not so with cast iron or vitrified clay pipe. Both are joined by means of a bell and spigot, usually, and both are difficult to cut and there is little room for error. You could easily end up short by a few inches, and you cannot cut the bell end of either kind.

If this happens, you will have to open the joints and shorten a few of the preceding lengths of pipe so the piece can be at least 6 inches long. By measuring carefully with a steel tape from the end of the house main drain to the point where the tank inlet will lie, you can determine if that is the correct distance, or vary it a few inches as necessary. Or, you can make a "dry" pipe run to see where the last section falls, and make whatever adjustments are necessary.

THE LEACH FIELD

With the septic tank placed, you can now add the proposed leach field to the map in its optimum location. The leach field must be down-slope from the house and the septic tank for proper drainage. The best location for it is on the south side of the house or other buildings if it must be close. Otherwise, it is best placed in any sunny location, preferably south-facing, and away from any shade trees or shrubs. Keep clear of walks, drives, easements, shady areas, and obvious obstacles. The field can be placed beneath a lawn, but should not have a vegetable garden planted over it (flowers are okay), nor should trees or bushes be near it. The entire field area should be at least 60 feet from any water well; 100 feet or more may be required. The overall leach field area must also be large enough to allow easy access and ample working room for excavating machinery and the piles of dirt that will be generated.

The leach field should not be located in an area that is continually or even periodically damp or marshy, or subject to occasional moisture buildup or flooding from storm runoff or snowmelt. If there is no other choice for a location, the area must be recontoured or otherwise drained off and reclaimed as dry land. In some installations, the ground is substantially built up to a higher grade level, using permeable materials, and the leach field placed within the new stratum.

Leach field configuration

The total length of leach pipe in the field is computed on the basis of liquid flow per time period and the rate at which the soil can absorb the liquid and also dispose of it by transpiration and evaporation. The data needed for this determination is developed through soil testing and finding the percolation rate, discussed in Chapter 2.

The design of the field is influenced by the required leach pipeline length, the topography, physical obstructions, the size and shape of the building site, and similar factors. Whether the leach line can be laid straight, must form a serpentine snake, divide into a number of arms, or take some other shape, will affect the overall space the field will occupy. Also to be considered is whether the leach lines must be set deep in a seepage pit, shallower in a seepage bed, or shallower yet in conventional leach trenches. Some general rules apply.

In a trenched arrangement, the distance from the outer wall of one trench to the next should equal 1½ times the effective depth of the trench. This is the distance from the center of the pipe to the bottom of the trench.

On slopes, the effluent must be prevented from "running away," or flowing rapidly out of the higher pipes to flood the lower ones. If this occurs, the lower pipes soon become clogged while the upper ones remain relatively unused.

The field should not be positioned where an automobile or truck might accidentally run over it, which could result in crushed pipes.

The field must be set back or separated a minimum distance to satisfy good neighbor policy and the local building department.

Finding the leach field size

Once you have run the percolation tests a sufficient number of times to secure relatively stable readings, you can be reasonably sure of the percolation rate in your proposed leach field area. Consistent readings from repeated tests indicate that the soil has become stabilized. It has absorbed a quantity of water that more or less imitates actual operating conditions. You can now use this time figure to find the leach pipeline length you need by referring to TABLE 6-1 and making the calculations as follows.

6-1. Time to Drain vs Required Drainpipe Wall Area

Rate	*Recommended minimum drain pipe*
1 minute or less	1.0 sq ft/gal/day*
2	1.5
3	2.0
4	2.5
5	3.0
10	4.0
20	4.6
30	5.2
40	7.0
60	9.–
unsuited	———

*This figure is for the "active" side of the trench only. It does not include the bottom of the trench nor the area above the center line of the drain pipe, and it includes only one side of the two walls of a trench. If you have a trench 10 feet long and the pipe was positioned with its center line just 1 foot above the bottom, that trench would have 20 square feet of active wall area or surface. Note: You might find this table at variance with other published tables on the same subject. These figures are a compromise.

These examples result in minimum figures. Increasing the length and thus the capacity of the system will provide a measure of safety against errors and a margin for future increased use. No harm will be done except, perhaps, to one's pocketbook.

Knowing the percolation rate provides you with the leach pipe trench wall area factor. Having determined the best depth for the leach line in your locale, you know how far down the centerline of the pipeline will be from grade level. Now you have to decide how much farther you want to dig. Remember, the effective wall area of the trench extends only from the centerline of the pipe (FIG. 6-5).

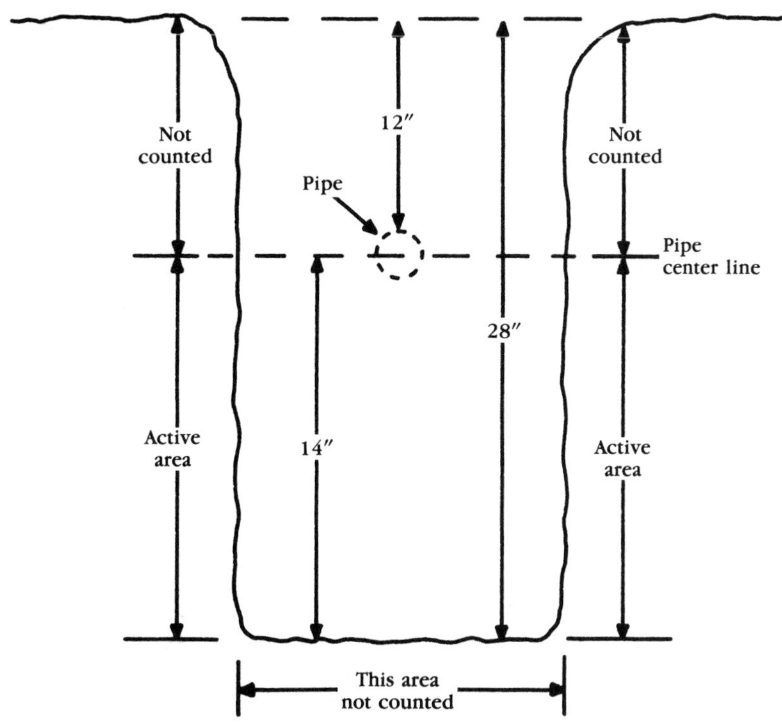

6-5 Measuring the active (useful) area of the walls of a leach trench. The active area in this example is 2 times 14 inches for every foot of trench length, or 2.4 square feet per foot.

Assume that you decide to make your trench 28 inches deep and position the top of the pipe 12 inches below grade. This leaves a side-wall effective distance of 14 inches. To find the active area of this trench per lineal foot, multiply:

$$14 \times 12 \times 2 = 336 \text{ square inches}$$

$$336 \div 144 = 2.4 \text{ square feet/lineal flow}$$

Assume that the daily effluent outflow averages 650 gal/day. To find the required wall area in square feet, divide:

$$640 \div 5.2 = 125 \text{ square feet (area for 30-minute rate)}$$

To find how many lineal feet of trench are needed, divide:

$$125 \div 2.4 = 52 \text{ lineal feet of trench.}$$

Let's take another example (FIG. 6-6). Assume the percolation rate shows 10 minutes for 1 inch of water level drop. That time indicates 4 square feet of trench wall surface is required. Suppose the effluent figure of 650 remains the same. Now instead of making the trench 28 inches deep, you make it 32 but keep the pipe top at 12 inches below grade. That leaves 18 inches of active wall surface.

$$18 \times 12 \times 2 = 432 \text{ square inches}$$

$$432 \div 144 = \text{square feet per lineal foot of trench} = 163$$

650 gal/day ÷ 4.0 (the 10-minute rate)

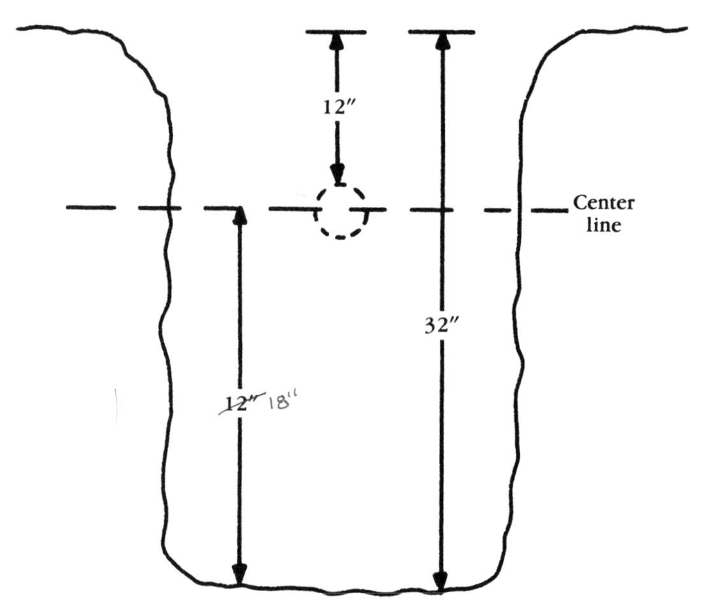

6-6 In this example, the active area is only 12 inches high. This makes a big difference in the total trench length required.

To find how many lineal feet of this trench you need, divide:

$$163 \div 4.4 = 37 \text{ feet}$$

As you can see, the perc rate of the soil plus the depth of the trench below the leach pipe center line taken together have an enormous effect on the total leach line required. In the first example you needed 52 lineal feet of pipe and trench. In the second, only 37 feet is needed. The difference is due in part to the faster absorption rate (10 minutes instead of 30), and the deeper trench (32 inches instead of 28) helped shorten the leach line.

The area comprising the bottom of the leach trench is not included in the absorption computations because it is the first area to become clogged with bacterial growth and debris. While this area will participate in effluent absorption for the first few months of system operation, the area will gradually lose its ability to absorb effluent and become useless.

The area above the center line of the leach pipeline is not considered in the absorption calculations because the effluent rarely if ever rises above this height in the trench.

Other methods

At least four other methods can be used to calculate the amount of leaching area required. One specifies that the leaching trenches must be 24 inches wide and 24 to 36 inches deep. A certain square footage of trench surface area is required according to the perc rate.

Another guideline assumes a trench width of 12 to 24 inches and a depth of 24 inches. The linear footage of leach pipeline is then determined by the number of occupants of the house and the amount of time required for 1 inch of water to fall in the perc test holes, in minutes (TABLE 6-2).

6-2. Leach Trench Length Required per Occupant

Occupants	*Minutes required per 1″ of water drop*		
	0–3	3–5	5–30
1–4	100′	150′	250′
5–9	200′	350′	700′
10–14	340′	500′	1000′

An older method specifies that the trenches must be 12 to 18 inches wide and 24 to 36 inches deep, a minimum of two trenches, no more than 100 feet of pipe in one trench, no less that 160 feet of total trench, and all trenches to be 6 feet apart. The total pipeline length is then based upon length in so many feet per bedroom, in concert with the perc rate in inches of water level drop per hour (TABLE 6-3).

A fourth method does not specify any particular absorption trench arrangement; in fact, trenches are not always used. Instead, it uses seepage area in square feet, calculated in relation to the perc rate in inches of level drop per hour, per bedroom (TABLE 6-4).

Remember that these figures are all minimums. No harm will be done even if you double them. In most cases, a larger installation does not cost much more than a minimum one, and it affords a better service life, a little insurance, and extra peace of mind. You will find, too, that in places where septic system installations are regulated, there will be a minimum leach pipeline or area requirement. A length of 100 feet of pipe or trench is typical, as is 150 square feet per bedroom or occupant.

6-3. Leach Trench Required per Bedroom

Water level drop (inches/hour)	*Trench length per bedroom*
12+	77′
12	83′
6	110′
4	127′
2	166′
1½	200′
1	220′
less than 1	unacceptable

6-4. Seepage Area Required

Perc rate (in./hr.)	*Area per bedroom (sq. ft.)*
less than 1	unsuitable
1	325
2	250
3	210
4	190
5	175
6	160
7	150
over 7	140

Leach field arrangements

Leach fields can be laid out in many ways, with the exact dimensions depending on factors discussed in this chapter. The following common arrangements can be installed as is, or modified to suit local conditions as necessary or desirable. One point to keep in mind, however, is that the more angles there are in the pipeline, the more the effluent flow will be impeded. Straight is best.

The leach pipe proper does not start right at the septic tank outlet. The line starts with a length of pipe, often only 5 feet, that has a solid wall. Keeping this section short is a good idea, but if conditions warrant, it can run out 20 or 30 feet or more. This leader may be coupled directly to the leach pipeline of perforated or separated lengths of pipe, or to a solid-walled manifold set at right angles and from which two or more leach lines extend like the tines of a fork, or to a distribution box. The box channels effluent flow in more or less equal measure into two or more leach lines. This is an important addition to the system when there are several equilateral leach branches or the lines stair-step downgrade fairly steeply.

A straight alignment is just that—right out of the septic tank and straight ahead for the total required length of the leach line. An otherwise straight alignment is sometimes cocked off in one direction or another with a single angle fitting, in order to take advantage of the topography or avoid an obstruction (FIG. 6-7). Approximately following a contour line in order to keep the pipe at a uniform depth below grade is a case in point.

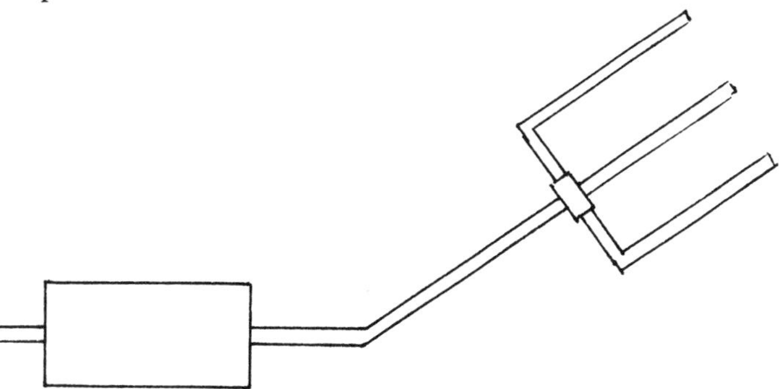

6-7 A straight leach line can be set at an angle to the tank to take advantage of ground contours.

Because the joints between leach pipe lengths need not be made up solidly, and in fact are often not made at all, a single leach line can be made to follow an arc, or snake gently back and forth through its course. Again, the reason may be to follow a grade contour, and effluent flow is not impeded appreciably.

Often it is necessary to run two or more leach branches in order to make the field more compact. If the entire field is flat, this is sometimes accomplished by laying a manifold or headpipe crosswise to the leader from the septic tank, then extending the branches outward at right angles to the headpipe. An alternative is to install a distribution box in place of the headpipe (FIG. 6-8). In both cases, only the branches are perforated or open-joint pipe, while the laterals are solid-walled and coupled.

A variation on this theme is to run a solid-walled leader pipe straight out of the septic tank. The branches tee off to left or right in parallel trenches. This is usually done in a level field, but can be advantageous on a slope. The leader follows the fall line of the slope at whatever angle is necessary. At each point where a branch joins, a distribution or diversion box is installed to equalize the flow of effluent. The branches run off at right angles to the fall line, nearly level (FIG. 6-9).

Another arrangement that has been used successfully on slopes is to lay the pipeline in switchbacks, with the lines more or less parallel and

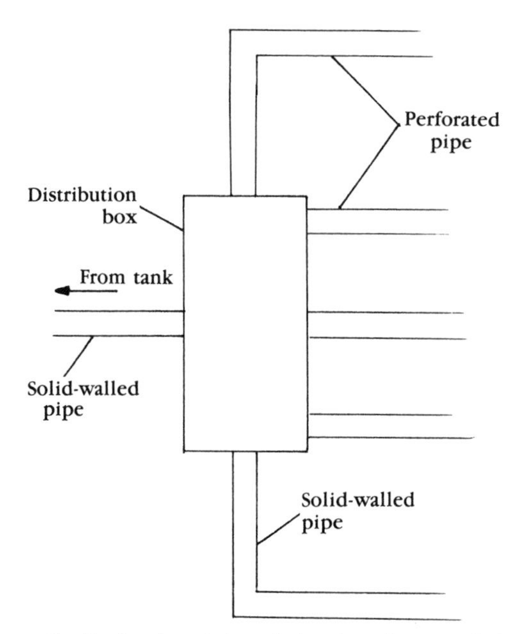

6-8 Sometimes a distribution box is installed to handle several short, parallel leach pipes.

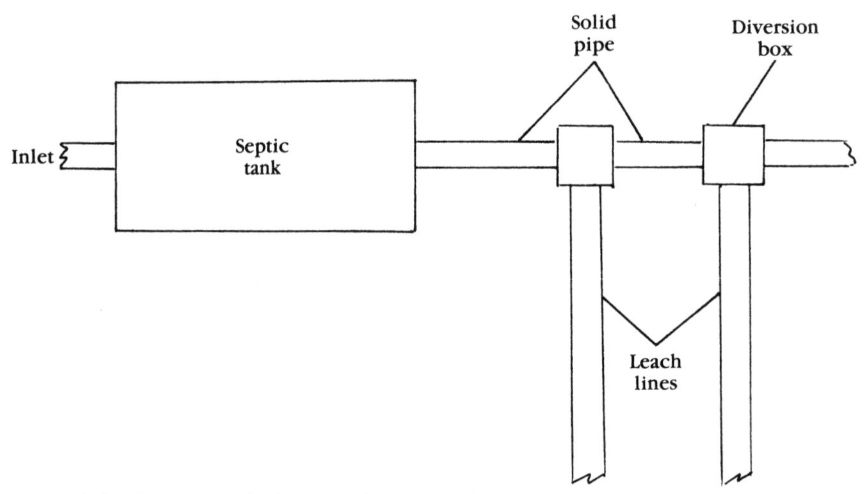

6-9 A leader pipe with diversion boxes can be set to a slope with leach branches extending outward at right angles, across the fall line of the slope.

each one lower than the previous one, laid level and running on contour across the fall line of the slope. The lines are joined with elbows and solid-walled and coupled pipe, first at one end and then the other to form a continuous line.

Sometimes the leach field must lie at a considerably lower elevation than the septic tank. In that case, a leader is run out, pitched at ¼ inch to the running foot, to an optimum point where a 45-degree fitting can be attached. The line falls at that angle until it reaches a point where another 45-degree fitting can be attached to bring the line level again and into the leach field (FIG. 6-10).

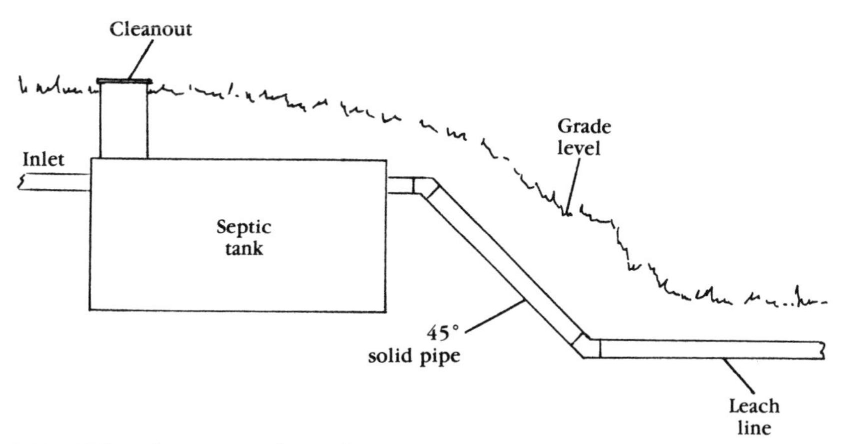

6-10 When the septic tank is well above the leach field, the leader to the leach lines is set on a 45-degree down-angle.

A large percentage of leach fields are installed trench-fashion, with the trenches parallel and anywhere from 2 to 8 or 10 feet apart, wall to wall. Sometimes, however, this is impractical and building a seepage bed (FIG. 6-11) is easier or will result in a more efficient field. Instead of

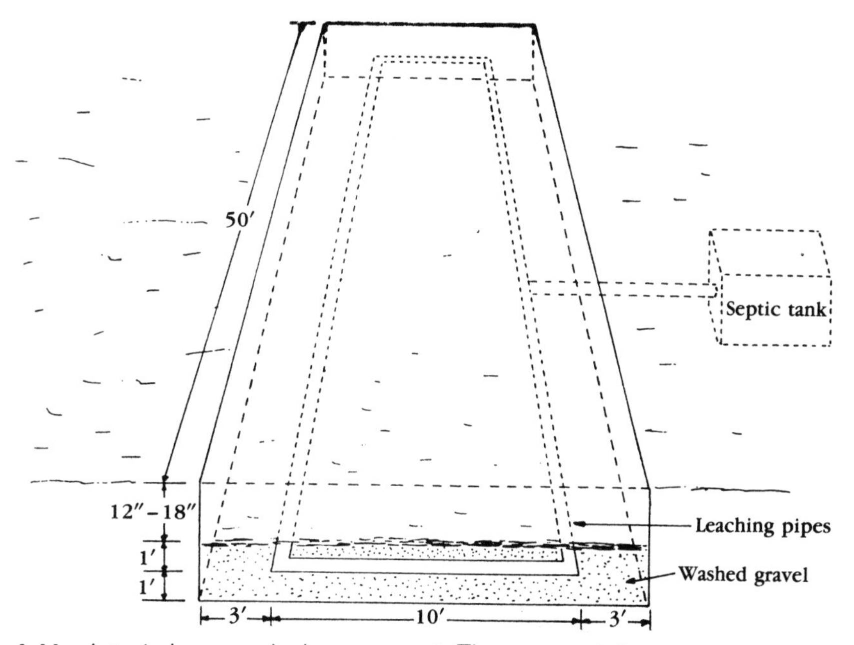

6-11 A typical seepage bed arrangement. There are variations.

trenches, a substantial rectangle is entirely excavated, perhaps 50 feet long and wide enough to accommodate several leach lines at their prescribed separation. The pipes lie anywhere from 2 to 3 feet below grade, usually anywhere from 4 to 10 feet apart, and the entire bed is filled with gravel topped with a foot of fill dirt.

A deep-bed, or seepage pit, is similar but buried much deeper (FIG. 6-12). The purpose is to set the leach lines below a surface layer of soil that is virtually impervious to moisture movement—no perc rate to speak of—and into a layer that is absorptive. In this case no evaporation or transpiration can take place, so the perc rate at the bottom of the pit must be good—at least 2 to 3 inches per hour. In this arrangement the leader is brought from the septic tank and dropped down at a 45-degree angle to a branched series of leach lines. They may be set like the tines of a fork, each capped at the open end, or they may be set in a closed grid with all pipes perforated or open-jointed. The pit can be as deep as necessary, even 15 or 20 feet, and the leach lines are set midway in several feet of gravel. Two or more vent lines may extend above grade to provide at least a small amount of oxygen to the aerobic bacteria.

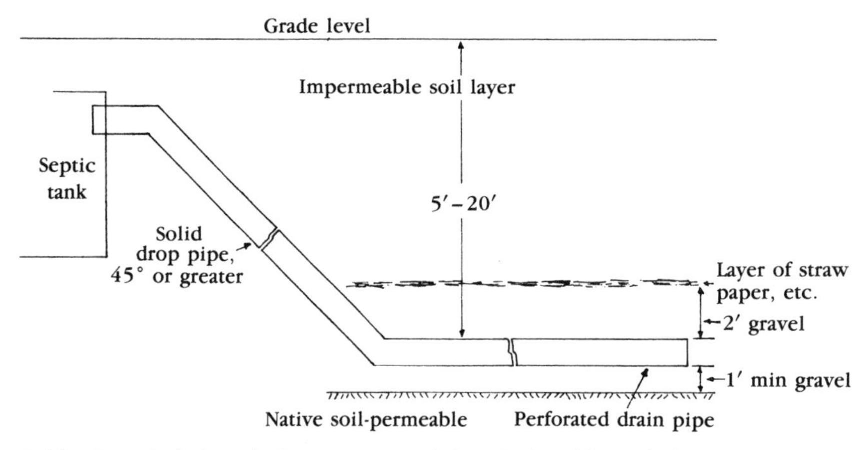

6-12 A typical deep bed or seepage pit installation. Many design variations are possible, such as using a porous concrete block lining in the pit.

The situation is just the opposite in grade-level or above-grade leach beds. Both are used in circumstances where the ground is or periodically can become damp or boggy, where drainage is marginal, where spring runoff or flooding might be a problem, where the water table is constantly high, and to combat similar difficulties.

In a grade-level bed (FIG. 6-13), the leach piping arrangement is designed to be as compact as is practical. The leach field is prepared by scraping it free of all vegetation and topsoil, with the grade leveled and taken down only a few inches. The area is then overlaid to a few inches above the surrounding grade with a permeable graded aggregate such as ¾-inch washed gravel (requirements vary depending upon place and conditions). The pipes are assembled on the surface of the aggregate,

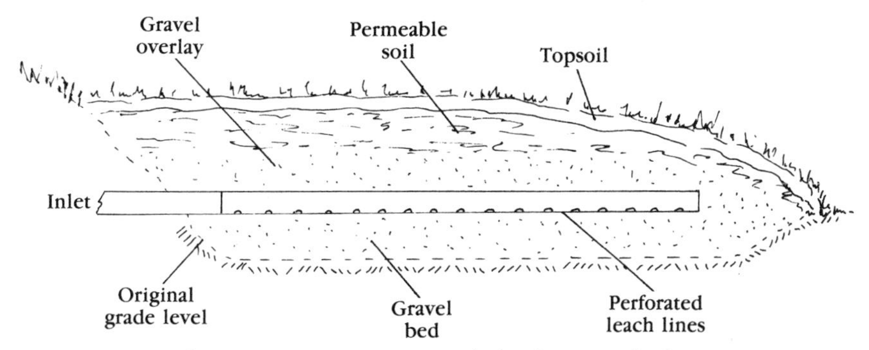

6-13 One way to construct a grade-level seepage bed.

covered with a layer of similar aggregate, topped with permeable soil, and finished, if need be, with a relatively thin layer of topsoil.

The above-grade system is similar, and is sometimes built where drainage is very problematical most of the time. In this arrangement the leach piping is typically a closed circuit of several parallel lines connected across both ends with cross pipes. A thick layer of permeable soil is laid on grade, after vegetation and topsoil has been removed, that extends outward well beyond the pipes. This is in turn covered with a layer of gravel. The pipe grid is set on the gravel and covered with several inches of the same material. The whole field is then covered with 2 feet or more of permeable but reasonably stable soil, such as graded sand mixed with a modest amount of clayey or silty soil, or perhaps road-base or crushed bank run gravel. This is covered with topsoil, creating a large mound that can be seeded to grass. The mound can be contoured and faired in around the edges to blend in with the surroundings, rather than looking like a big burial site.

Some recent innovations in the way leach lines are installed, too, are not yet in common use, but can answer some of the marginal-drainage problems.

One such arrangement involves encasing the leach pipes in a special plastic clamshell-like covering as it is laid. A wider than usual trench is required—3 feet or more. The trench bottom is leveled and a thin layer of gravel put down. The bottom half of the pipe casing, which looks like an extra-long, overgrown banana-split dish, is bedded on the gravel and the leach pipe set into it. Then the case is filled with more gravel, covering the pipe, and the top half of the case is snapped in place. The whole affair is then covered with yet more gravel and the trench backfilled to grade level with spoil dirt.

Another method makes use of a series of large, bottomless and topless cast concrete boxes. The boxes are set in ranks, typically in four rows of three each. They are then partly filled with gravel. The leach pipes run through the boxes, and are covered with successive layers of gravel of varying sizes and gradings, creating a set of combination filtration and evaporation beds. The whole installation is covered with a permeable layer like straw or roofing felt, then backfilled to grade level or slightly above with clean spoil dirt.

Details of leach field arrangements and installations vary widely throughout the country, as do job conditions, and there are many more than have been discussed here. In addition, research continues in the sewage disposal field—this is one of the more pressing problems in many areas of the country—and new ideas, products, and recommendations applicable to individual residential septic systems appear regularly. Everything that has been discussed herein is time-tested and workable under appropriate conditions. However, you should make a point of checking out the latest information when you start to design your own system.

Chapter 7

Installing the system

Once your plans for the system are complete, you can begin installation. Usually the septic tank is set first, and the sewer line laid between the tank inlet and the outlet of the house main drain. Then the leach lines can be installed and the field completed. All of this work, except for the house main drain connection, can theoretically be done at any time, no matter what stage of construction the house happens to be in, but most builders prefer the job to be far enough along that the house main drain is already in place and ready for connection to the tank first. This minimizes the possibility of running into problems with placement, pipe pitch, and trenches.

EXCAVATING

In most installations, excavation is done with a backhoe or similar machine. The job is easier if the machine is fitted with a bucket of the same width required for the leach line trenches; width of the sewer line trench is not crucial and can be whatever is convenient. If a deep bed is part of the plan, often a hoe with an extra-long boom, or even a power shovel, is needed.

Sewer line

Usual practice is to start at the foundation wall of the building and dig a trench for the sewer pipe back to the septic tank location. This trench should be rough-graded, if possible, to a drop of about ¼ inch per running foot from house to tank. A width of about 2 feet is generally ample. The depth should have been preplanned, based upon the position of the house main drain stubbed through the foundation wall, the sewer line length and drop, and the required elevation of the septic tank inlet (FIG. 7-1).

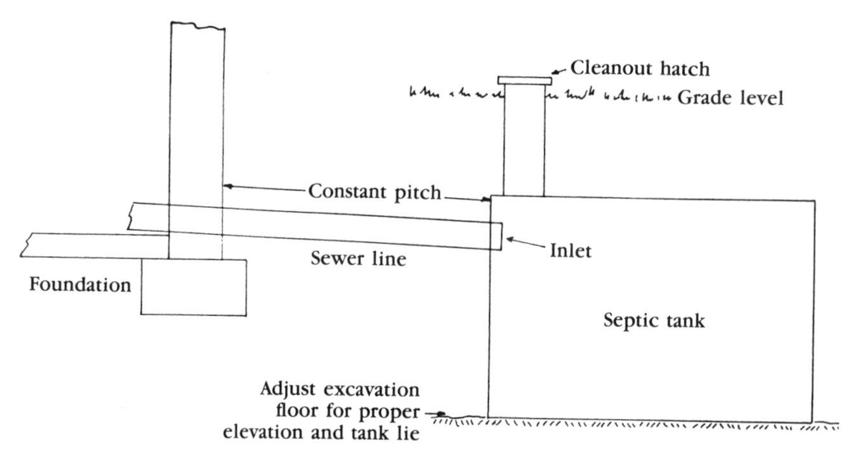

7-1 The elevation of the septic tank must be adjusted to exactly align with the house foundation, house main drain, sewer pipe inlet and outlet, and approximate final grade in the tank area.

Septic tank

Next, the tank hole can be dug, large enough for some working room around the outside of the tank, especially if it will be built in place. The bottom of the hole should be as flat as possible, and exactly at the required depth for the bottom of the tank so it will sit on undisturbed native soil. Again, this is not always possible. Depending upon soil conditions, some installers prefer to leave the floor of the hole a few inches high, then finish up by hand in order to control the depth and level exactly.

Leach lines

Finally, the trenches for the leach lines are dug. They should be kept uniform and properly spaced. Keep constant track of the depth, making the trench bottoms reasonably flat and level. The sidewall height will vary because of ground surface irregularities—the key is to maintain a level trench bottom so the leach pipes will also lie level without a lot of extra work. Digging too deep means more work and extra gravel, and too shallow means a lot of hand work to correct the problem. The specific dimensions of the trenches are determined during the system planning stages, depending upon site conditions and local regulations.

If a seepage bed is required, it probably will be more easily excavated by a backhoe with a bucket mounted on front, or perhaps a front-loader. Again, the depth should be well controlled—getting too deep is very easy—and at least a third, possibly half, of the spoil dirt can be hauled away or shunted off for recontouring elsewhere, as it is dug. It will be replaced by gravel.

In the case of a deep bed, the surface area of the excavation is typically small—perhaps 20 by 20 feet or so—but an enormous amount of spoil dirt comes up if the hole is very deep. As much as half can be trucked off immediately to get it out of the way. It is a good idea to dig about 1 foot or so too deep and scrape the bottom as clean of loose dirt as possible, using only the machine. No one should be allowed down into the excavation to do any handwork unless it is shored up on all sides, or relatively wide and shallow enough that being trapped by a collapsing wall is not a possibility.

SETTING THE TANK

Usual procedure is to set the septic tank first. The bottom of the excavation should be flat, and hard, undisturbed native soil. If this is not possible, a few inches of damp sand can be spread, leveled, and compacted, or a shallow layer of pea gravel can be laid in. The elevation of the hole bottom should be adjusted to compensate, so the tank inlet and outlet will be at their proper elevations.

If the tank will be site-built, construction can proceed as outlined in Chapter 4. A manufactured tank is typically set in place with a special boom truck. This arrangement slides the tank off the truck bed, suspends it over the hole, and lowers it into place. The same operation can be managed with a crane, backhoe or front loader, although not as easily.

There are three crucial elements in this procedure. The first is to make sure the tank inlet faces the house, not the leach field; more than one tank has been set backwards! The second is to make sure that both the inlet and the outlet are oriented correctly for their respective pipelines. The third is to make sure the tank sits dead level along the axis running across the inlet and outlet; the side-to-side level doesn't matter as much. In most tanks, for proper drainage, the inlet centerline lies about 2 inches above the outlet centerline. If the tank is off-level that relationship is changed, and trouble will result.

If the system calls for two tanks, they may be set end to end or side by side. Again, the inlet-outlet relationship is crucial. To ensure proper drainage into, between, and out of the tanks, the second in-line may be set a bit lower than the first. In any event, the flow path must run smoothly from higher to lower elevation.

In most cases the backfill can be placed as soon as the tank is set, up to a level slightly below the inlet and outlet. This makes working safer and easier. The installation should not be covered up, however, until it has been completed and checked out for leaks. In many areas an inspection is also required at that stage.

SEWER LINE

Unless local regulations dictate otherwise, the material of choice nowadays for sewer line in single-family residential installations is 4-inch plastic pipe. It must be solid-walled and tight-coupled. The line should lie on a grade of ¼ inch per running foot for rapid (but not too rapid) drainage, and there are several ways to accomplish this.

The best arrangement, especially if the local soil is expansive, is to cut the trench so the bottom will lie about 8 inches below the projected bottom location of the sewer pipe. Attach one 2 by 2 pointed stake (each about 12 or 14 inches long) close to each end of an 8- or 10-foot 2 by 6 (FIG. 7-2). The 2 by 6 should be either heartwood redwood or, much better, wood pressure preservative treated to 0.60 pounds of preservative salts per cubic foot (wood foundation material).

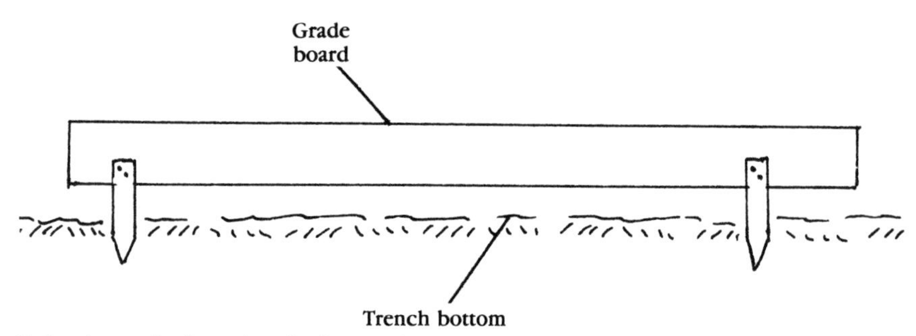

7-2 A grade board staked into the trench bottom assures ample sewer pipe support, constant pitch of line, and no sagging.

Alternating stakes, drive each into the ground a bit at a time so the centerline along the upper edge of the 2 by 6 coincides with the alignment of the sewer pipeline. Stretch a taut line from the bottom of the house main drain stub to the bottom of the septic tank inlet as a reference guideline. If you have set the tank correctly, the string should be on a ¼-inch per foot downgrade. Drive the stakes until the top edge of the 2 by 6 lies even with the taut line (FIG. 7-3). Remove the taut line and fill all around the 2 by 6 with clean, fine earth fill, pea gravel, or damp sand.

Finally, lay the pipe on top of the 2 by 6, making tight couplings as you do so. An adapter fitting will probably be needed at the house end, perhaps at the tank as well. Often the tank end is made up by running the tee fitting outward from inside the tank to make up with the end of

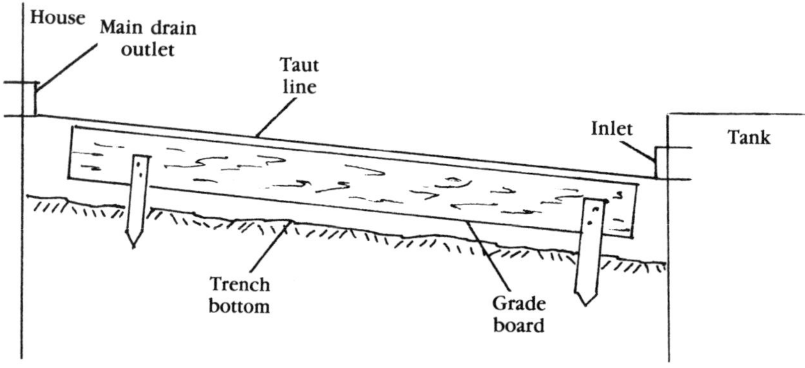

7-3 Use a taut line to maintain the correct sewer pipe pitch when you install it.

the sewer line. Caulk thoroughly around the inlet fitting and make sure it is well sealed.

An alternative way to lay the sewer pipe is simply to grade the trench bottom, add a layer of sand or pea gravel (sometimes only rough fill dirt is used), and bed the pipe in it. This is less satisfactory because the pipe can get off-grade during installation, or be damaged during backfilling through lack of support. Also, if the soil is expansive or susceptible to frost-heaving, the pipe can sag, hump, or buckle, even break or pull free from the tank inlet, causing blockage and all sorts of problems.

THE LEACH FIELD

The leach field can be installed any time that is convenient, even before the tank is set in some cases. The field may consist of one or more trenches, a shallow bed or a deep bed, plus a short leader pipe to the leach lines proper. There may or may not be a distribution box or one or more diversion boxes.

Leader

The pipe running from the septic tank outlet to the distribution box, the first diversion box or the first length of leach pipe should be solid-walled and tight-coupled. It can run straight out from the tank, or be angled out or angled down, or both. Although length is not a critical factor, shorter is better.

This line will carry liquid that contains few solids, and those are (or should be) only particles. Thus, a fairly steep pitch to carry solids is not required, as it is with the sewer pipe. The line should be pitched downward from the tank however, and ⅛ inch per running foot is generally considered adequate. Some installers prefer to use a ¼-inch pitch, and this does no harm. What makes the best pitch should be given some consideration, depending upon the layout of the leach lines and presence or absence of distribution or diversion boxes.

The theory is that you want the effluent to run from one end to the other of the entire leach line, or set of lines, in an even flow. You don't want the forward parts of the line to fill and leave the farther parts dry, nor do you want a high percentage of the effluent rushing into the far end of the line. You do want the effluent to equalize and drain for about the same length of time throughout the entire system.

The best method of installing the leader pipe is the same as that for the sewer pipe. Base the pipe on a properly pitched grade board staked into undisturbed native soil. Bed the pipe in gravel. The same that will be used in the leach field, typically ¾-inch washed gravel, will work fine, but so will other clean fills. This is the installer's choice.

Distribution box

All that is required to install the distribution box is that it be properly oriented so the leach lines head off in the right directions and the box should be level or pitched just slightly toward the outlets. It should be bedded in sand, gravel, or clean fill so frost-heaving is minimized and leveling is easy. At least 1 foot of earth should cover the box. Diversion boxes are installed in the same way.

Trenched leach lines

The general procedure for installing leach lines in trenches is about the same everywhere, although many of the details will vary. Distance between trench walls, trench depth and width, the depth of gravel under and over the pipes, whether or not a filter layer is laid over the gravel, the kind of pipe required, and similar matters depend upon local regulations and are settled during the planning stages.

One point of concern is the pitch of the leach lines. As mentioned earlier, the object is to have the effluent spread out about equally, in both distance and time, through the whole field (FIG. 7-4). Some author-

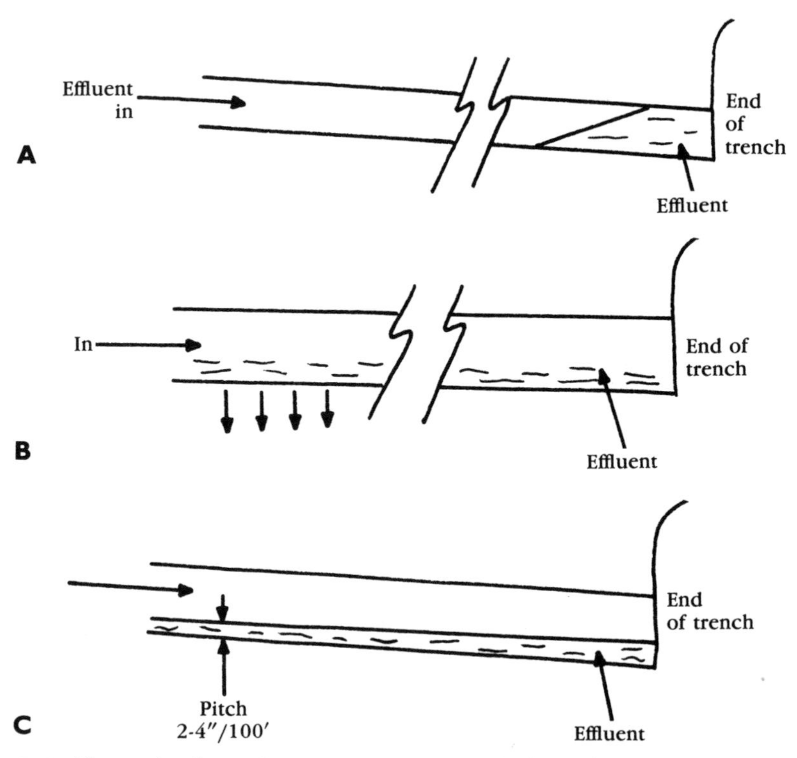

7-4 The pitch of a perforated drain pipe influences effluent distribution and leach field life. A. The pipe is pitched too steeply. The effluent all runs to one end. As a result, this portion of the pipe is overloaded and soon becomes plugged. B. Here the pipe is level. The effluent divides evenly at first, but because it spends more time at the head of the pipeline this portion is overloaded and plugs up first. C. Here the pitch is gentle. A little more effluent flows toward the end of the pipeline than remains along the head end, but that results in relatively even use by both pipe and surrounding soil.

ities insist that the pipes all be laid dead level. Others want the lines laid at a downward pitch from the leader pipe or distribution box at a rate of 2 to 4 inches per 100 feet of pipe. When plastic pipe is being laid in the commonly available 10-foot lengths, a figure of ¼ to ⅜ inch of drop per pipe section is sometimes used. If you have a choice, opt for a greater pitch in highly permeable soil; reduce the pitch correspondingly in soils of lower permeability.

In a typical installation, the first step is to clean any loose dirt out of the trenches by hand. The pipes can be bedded in gravel and supported only by grade stakes, or they can be set on continuous grade boards as discussed for the sewer pipe (a necessary arrangement for clay drain tile). The boards used for this purpose are usually untreated pine or fir 1 by 6s, which eventually rot away.

For the former arrangement, mark off 5-foot intervals and drive a grade stake at each point. The tops should be flat and aligned. If you use grade boards, proceed the same as for the sewer pipe. For a level installation, simply drive the stakes until all the tops, or the upper edges of the grade boards, are dead level and at equal elevations in the trenches (or proper elevations if the trenches vary in elevation, as in a sloped field).

For pitched installation, select a 10-foot, straight edged board and drive a 4d nail in one edge close to the end, leaving the head sticking up ¼ to ⅜ inch. Set the first grade stake or grade board end at the starting elevation. Rest the nail head on this point and the board end on the next downslope point. Place a spirit level on the upper edge of the board. Drive the downslope stake in. When the bubble is centered, the grade stake or board is set to the desired pitch of ¼ to ⅜ inch per 10 feet (FIG. 7-5).

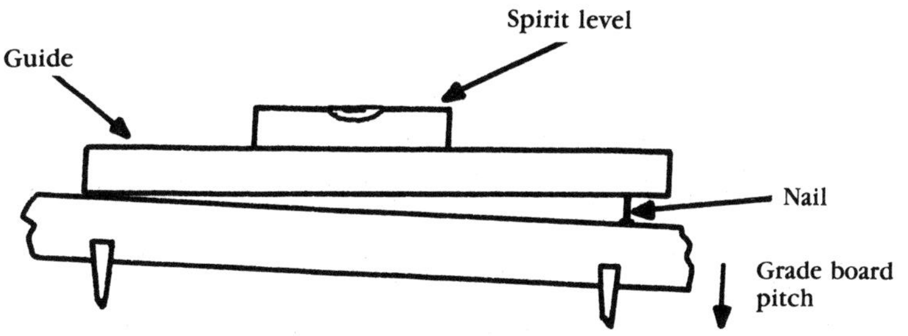

7-5 A straight-edged board, a nail, and a spirit level make pitching grade boards a simple matter.

Once all the grade stakes or boards are correctly set, fill the trenches with the required type of gravel. In most installations the depth of this gravel bed is 12 inches, and sometimes it is more. In the absence of other specifications, use ¾-inch washed gravel. Rake the gravel out smooth and level, even with the tops of the grade boards or stakes.

Lay the pipes in place and nest the sections in the gravel atop the grade boards or stakes. Use perforated plastic pipe here, with the perforation placed downward, or clay drain tile sections about ¼-inch apart.

No couplings are needed on plastic pipe, but they are often slipped on dry (no welding solvent) to help keep the sections aligned. Open joints should be covered over with a flap of roofing felt or a similar material to keep gravel out.

Cover the pipes with 4 to 6 inches of the same gravel that was used for the bed. Place the material by hand, taking care not to cause any damage or move the pipes. Cover the gravel with an infiltration barrier to keep soil from sifting down into the gravel bed. This layer is often 15- or 30-pound roofing felt, untreated building paper, or even 2 or 3 inches of straw (hay is not a good choice). There are also special materials available made for the purpose. Whatever is used should be permeable, not a waterproof vapor barrier. Finally, backfill the trenches with clean fill, and mound the loose dirt up about 6 inches over the trench. In due course it will settle down level. Figure 7-6 shows a cross-section of a completed leach trench.

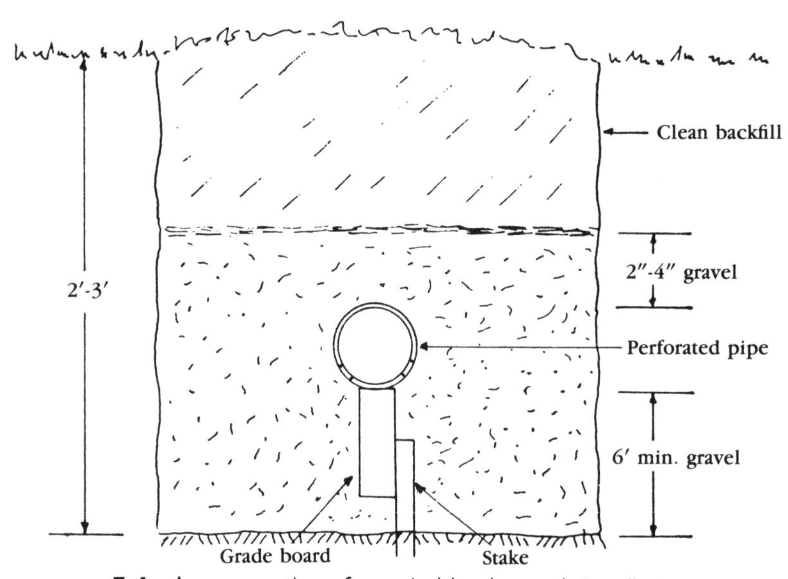

7-6 A cross-section of a typical leach trench installation.

Some trench layouts consist of two or more connected runs set in tiers that cross the fall line of a slope at right angles. The pipelines are connected at one end to a leader that follows the fall line. Despite the different design of the field, the installation is made almost the same as for a straight-line trench layout. The leach lines are laid level or just slightly pitched, bedded in gravel in the usual way. The only differences are in the pitch of the leader pipe, and the possible presence of diversion boxes to split up the effluent flow more or less equally to all the lines. Fittings in the leader pipe are often solidly made up so they cannot slip apart.

Many installations are now required to have vent pipes in the leach field. Usually these are simply an extension of each individual leach pipe line (FIG. 7-7). Connect a 90-degree elbow to the end of each line instead of a cap, and continue the pipe line straight upward with a solid-walled length that reaches about 12 to 18 inches above grade. These pipes are often left open, but a better arrangement is to add a pair of elbows to make a shepherd's crook on each one, or at least fit screened caps. For a less obtrusive appearance, make these vents of black plastic pipe (much of the pipe used for leach lines is bright white).

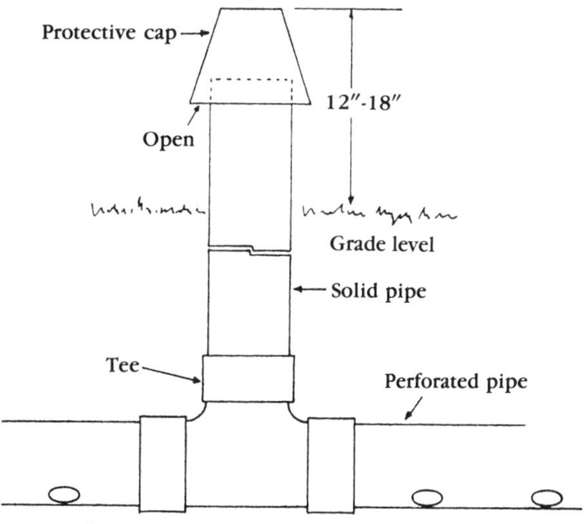

7-7 One way to add a vent pipe to a leach line.

Seepage bed

A seepage bed is typically laid out as a rectangle. The entire area is excavated and two or more parallel lines of leach pipes are laid on a gravel bed (FIG. 7-8). The overall arrangement is just about the same as a leaching trench set-up minus the trenches. Construction starts by setting up grade stakes or boards, pitched or level as required. Then the gravel bed is put in place and the pipes set, the gravel cover layer is raked in, and the bed is backfilled.

Seepage pit or deep bed

In a seepage pit or deep bed arrangement (FIG. 7-9), usually a couple of feet of gravel graded from ½ to 2 inches or so in size are first laid in the bottom of the excavation. In some circumstances, more may be used. With the sidewalls shored up, the gravel should be leveled and raked out smooth.

The leach piping can then be laid out on the gravel, either level or pitched slightly down and away from the leader pipe. Often as not the arrangement is in the form of a closed grid composed of half a dozen or more pipes with a manifold pipe at the head and a cross pipe at the foot, all interconnected with elbow and tee fittings. The center of the mani-

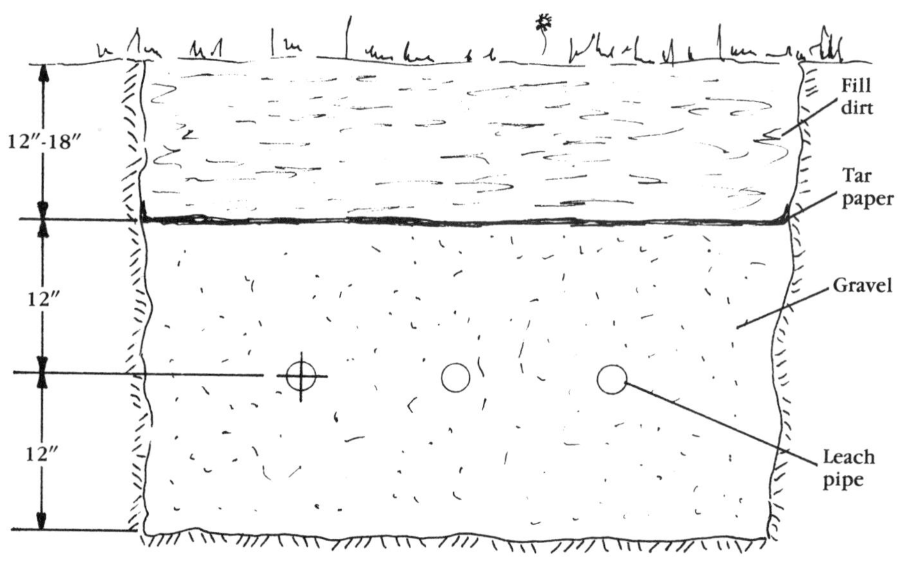

7-8 A cross-section of a typical seepage bed installation.

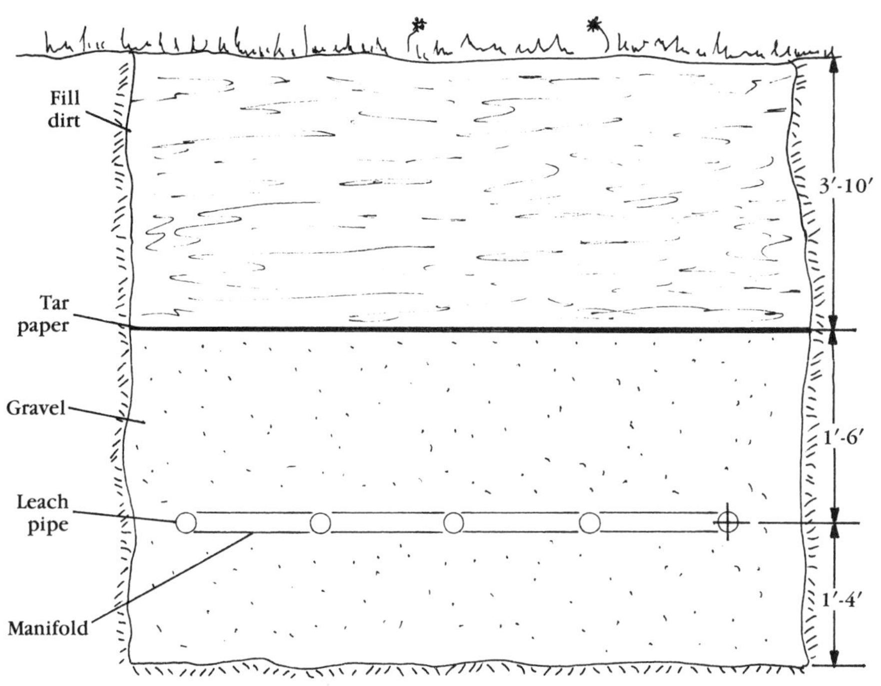

7-9 A cross-section of a typical deep bed installation.

fold is usually the connecting point for the leader pipe coming from the septic tank, angled downward at 45 degrees.

When the piping is in place, it can be stabilized by surrounding and then covering the lines with gravel. This should be done carefully by hand to prevent any damage or shifting; use ¾-inch washed gravel for easy handling. Then a layer of more gravel can be gently dumped in place, generally about 1 to 2 feet deep, and raked out roughly level. The next step is to lay an infiltration barrier of roofing felt, building paper, or straw over the gravel. Then the pit can be backfilled, using only clean fill that is free of any rocks larger than fist-size. If that is not readily available, use pit-run gravel or something similar up to within 6 inches or so of grade level, and fill the rest of the way with topsoil.

Sometimes two or more air vent pipes are included in the piping design. These extend upward from tees inserted in the leach grid. They are simply inserted in the tee fittings after the grid has been arranged and anchored, then filled around with gravel and backfill as that process proceeds. Care should be taken that they remain vertical and do not become cocked or pull out of their fittings. Termination is usually in a shepherd's crook or screen cap set about 12 to 18 inches above the finished grade.

Part 2

Water wells

Chapter 8

Groundwater

When I was a boy, I spent time on a farm in the Catskills, where we got our drinking water from a spring that gushed from a crack in a hill. A hundred yards away there was a creek full of good, cold drinking water. Even today some houses get their water from shallow dug wells in the side yard, or pull water from nearby river-fed irrigation ditches that run along the town streets, or pumped from a pond or lake. In many parts of the West, neighbors get together and chip in to hire a professional driller to go down several hundred feet to find water.

The point is simply that in some places water is easy to find. In some places it may be pooled in a still body, run in a creek or river, or flow from the ground in the form of a spring or artesian well. In others, a lot of hard work and considerable time and equipment are required. If water is visible, there might be no problem. But if there is no water in sight, how can you tell whether or not you can reach water by putting down a well? How can you tell how deep you will have to go? How much water can you expect for your efforts?

Unfortunately, none of these questions can be answered with a formula or a set of instructions or measurements. To get answers with any useful degree of accuracy, you must understand the nature of groundwater and aquifers and study the problems the way a hydrologist does.

WATER IN THE EARTH

When you dig down into the earth, you pass through a layer that hydrologists call the *zone of aeration* (FIG. 8-1). This consists primarily of soil. The top of the zone is the surface of the earth and the bottom can be described easily, but not accurately, as resting on the water table.

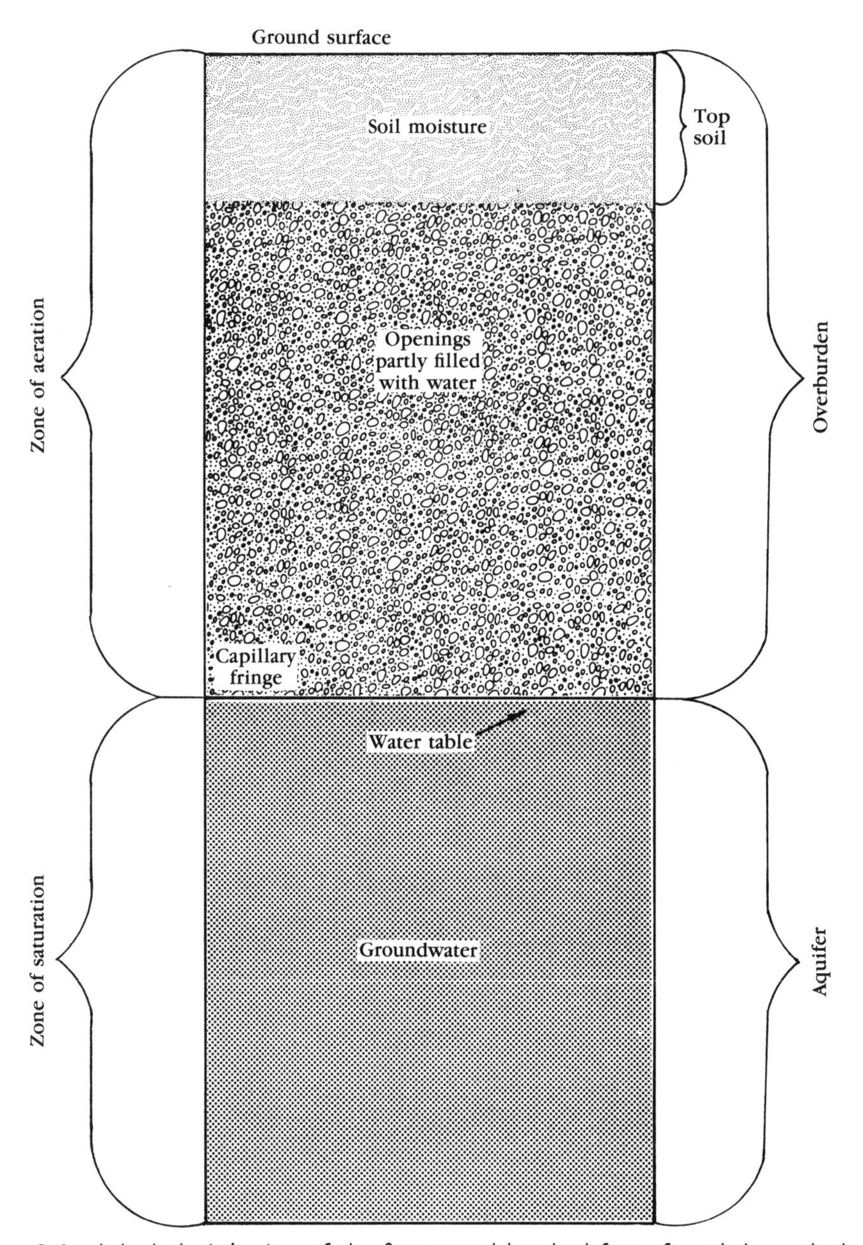

8-1 A hydrologist's view of the first several hundred feet of earth beneath the surface.

There is almost always a little water to be found in the zone of aeration, most of which is held in the smaller openings. The larger openings generally contain air. Following a heavy rainfall this zone becomes saturated with water. During dry spells the water is removed by surface evaporation and by plant transpiration.

The water in the zone of aeration will not enter a well. Digging a hole in moist or damp earth will not result in a flow of water into the hole.

Groundwater

Groundwater is found beneath the zone of aeration. The uppermost level of the groundwater is called the *water table*. Groundwater sometimes takes the form of an underground stream flowing slowly in a dark tunnel. It might even be in the form of a lake in a cave, but these are rarities. Almost all groundwater resides in tiny cracks and pores in the rocks beneath the aeration layer. Groundwater is found in fractured rock, sand, sandstone, and gravel.

When rain falls upon the earth some of it is absorbed by the zone of aeration, and the remainder runs off at the surface level or seeps, downward into the earth to join existing groundwater in the zone of saturation (refer to FIG. 8-1). The addition of more groundwater raises the level of the water table—the level of the water in the pores, cracks, and crannies in the rock or sand is raised.

Aquifers

The cracks extend only so far down, and no further. The fractured rock, sand, or gravel zones contain *aquifers*. This term is derived from two Latin words: *aqua*, meaning water, and *ferre*, to bring. Thus, an aquifer brings underground water; it is a waterbearing stratum of permeable rock, sand, or gravel.

A hole dug into an aquifer will turn into a water well because the openings between the pieces of sand or gravel or shattered stone are so large that the water molecules can easily flow.

The thickness of an aquifer from top to bottom varies with the nature of the earth. An aquifer can be only a few feet or inches thick, or it might be hundreds of feet thick. It might extend for miles without a break, like the Ogallala Aquifer which stretches from Nebraska to Texas. This aquifer is as large as California and contains as much water as Lake Huron. The aquifer might be only pond-sized, or it might stop abruptly because of some prehistoric geologic upheaval. Generally an aquifer follows the topography of the land. Where the land dips, the aquifer does likewise.

Groundwater generally flows downhill in the same direction as the overall downward slope of the ground surface. Its movement through the aquifer is slow as it meanders around barriers posed by solid rock. The water will follow the easiest routes through sand or gravel, and eventually might emerge as a spring or join a stream 50 miles or more away from the point of origin, having served numerous points along the way.

Water can travel through gravel at a rate of hundreds of feet a day. In fine sand or silt, the rate might be no more than a few inches a day.

While the flow of an open stream or river can be measured in tens, hundreds, or even thousands of cubic feet per second, the flow of groundwater is usually measured in terms of cubic feet per year.

There is no firm relationship between aquifers and the depth at which they exist. In New England, for example, very dense granite extends upward to or near the earth's surface. In the Great Plains, very porous sandstone lies several thousand feet beneath the surface. Generally, the farther down you go the less permeable and less porous are the rock formations you find. Fresh water has been found at a depth of 6000 feet and salt water, along with oil, at depths of more than 20,000 feet. But as a rule, few wells drilled deeper than 2000 feet find water. The enormous weight of the rock itself squeezes shut all the cracks, crevices, and crannies.

Groundwater is normally recharged or replaced by rainfall (FIG. 8-2). The percentage of the rain that replenishes the supply of groundwater depends upon the nature of the soil upon which the rain falls and its pitch. If the soil is porous and relatively flat, much of the rain will seep into the aquifer. If the soil is hard and has a high percentage of clay and, in addition, if the surface is steeply pitched, most of the rain will run off and not reach the aquifer.

For a well digger, this means that drilling a well at the bottom of a large hill does not guarantee finding a dependable source of water in adequate quantity. There might be an aquifer within easy drilling distance, but if the surrounding area does not permit ready recharging there will be little dependable water in the well. Instead of recharging the aquifer, the rain will come down the hill in a troublesome surface runoff stream. If the soil on the hill is porous but overlays a solid layer of clay, there will be less surface runoff but the recharge rate will not be much better.

WATER QUANTITY

The quantity of water that can be more or less continuously drawn from a well depends upon the permeability of the aquifer, its size, the recharge area, rainfall in the recharge area, and the length of the well screen extending into the aquifer. To a very small extent, the diameter of the well also has an effect. The larger the diameter, the more cracks the well intersects and the greater the quantity of water that will enter the well. But should the water table level drop below the bottom of the well, no water will be available no matter what its diameter.

All aquifers rest upon solid, watertight rock. The quantity of water that can be contained in an aquifer of a given size depends upon the porosity of the stratum. In this case, porosity equals the spaces, cracks, and crevices between the pieces of rock capable of being filled with water that cannot be held in place by capillary action.

In sand, the porosity can amount to 30 or 40 percent of the total volume. The porosity of sand and gravel depends not only upon the size of the pieces of stone but also on their relative sizes. When the grains are

A

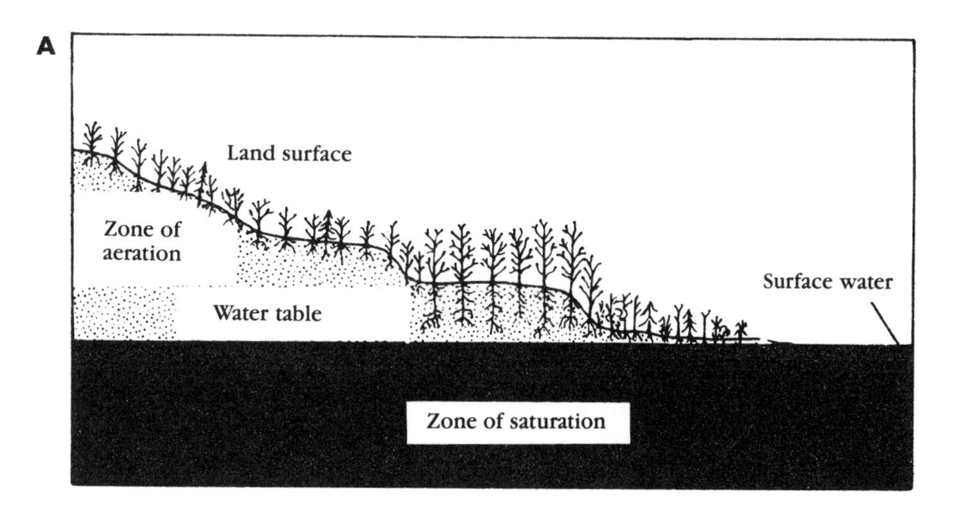

B

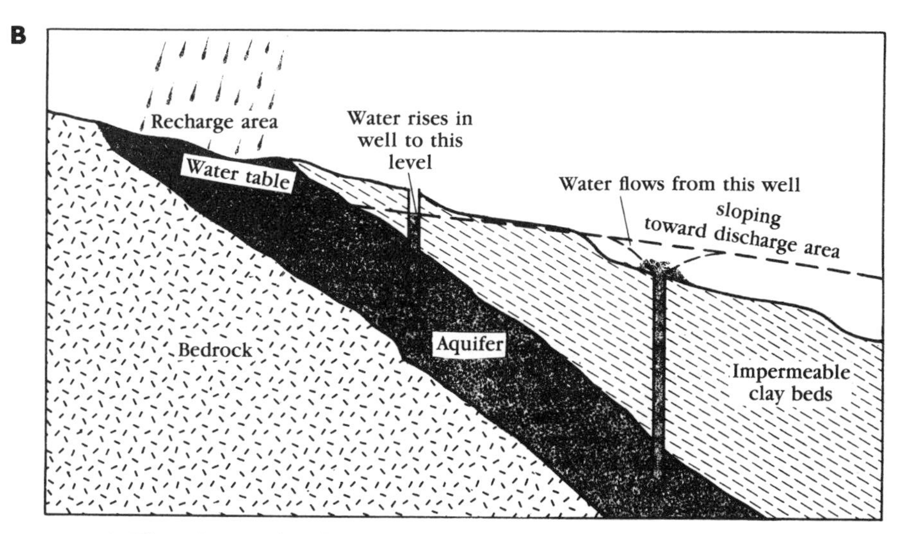

8-2 A. The relation of surface water to the water table. B. Rainfall can recharge an aquifer. Note how the water table appears in the normal well and in an artesian well (at right).

all about the same size, or well sorted, as geologists say, the grains do not pack as well as when the sizes vary and the grains are poorly sorted. Then the smaller grains slip into the spaces between the larger ones and the result is less air space and less porosity.

Different kinds of rock vary widely in porosity. In some very dense rock such as granite, porosity will be less than 1 percent. The other end of the scale reaches 40 percent for sand, gravel, and unconsolidated rock. In some instances, the absence of an appreciable quantity of pores does not stop the flow of underground water. Some compact, consolidated rock such as granite and slate will have sufficient cracks, called *joints* by geologists, to permit the flow of groundwater. When and where these joints intersect, water can flow almost as easily as through an ordinary water pipe. Wells are practical in such consolidated, dense rock only when there are many cracks and the well intersects a goodly number of them.

Groundwater moves much more rapidly through some kinds of rock than through others. Movement is free through sandstone because it has natural pore spaces. The flow through solid granite, schist, or slate is negligible. Clay and silt are almost permanent barriers to water movement, but coarse gravel offers almost free passage. Limestone is often cavernous, and water frequently flows through it faster than it does through other formations.

GROUNDWATER RESERVOIRS

Some geologists use the words *groundwater reservoir* and *aquifer* interchangeably, but generally the former term is used to describe the zone of saturation. Assume that a particular aquifer covers an area of 10 acres, and it is possible to measure the total height of the water in the aquifer from the surface of the solid underlying rock to the water table. This body of water, held in suspension in the cracks and pores of the aquifer, is the groundwater reservoir—the zone of saturation (shown in FIG. 8-1).

It is from all the groundwater reservoirs combined that we draw our well water. The total quantity runs into the billions of gallons each day. No one knows exactly how much water there is in our groundwater reservoirs, but the generally accepted estimate is that our underground water supply may be several times greater than all our fresh-water lakes and surface reservoirs combined. This sounds like a great deal of water, but in fact it is not—we are actually running short. This is not a diminution of the total amount of water on this planet, but rather of supplies of water that can readily be tapped and made potable when and where needed.

One reason for this shortage is usage. We are simply removing water faster than nature is replenishing it. Another lies in the makeup of a typical groundwater reservoir. When you have water in a lake or an artificial storage pond, all of that water is immediately available. You can remove the water as fast as mechanical pumping will allow. But underground water is contained in billions of minute pores and cracks. The rate of water flow from the far reaches of an aquifer to a well and pump is very limited. The rate at which you can draw water from an aquifer is therefore limited as well. Thus when you pump water from a well more rapidly than the groundwater can enter the same well, it will in due course run dry even though there may be more water in the aquifer.

Cone of depression

Because water does not flow freely through an aquifer, when water is drawn from it by means of a well the water table (water level) in the immediate vicinity falls. If you draw water from a pond the entire surface level drops, but this is not true of an aquifer. As water is pulled from a well only the local water level drops variably within a certain area, and this area is called a *cone of depression* (FIG. 8-3).

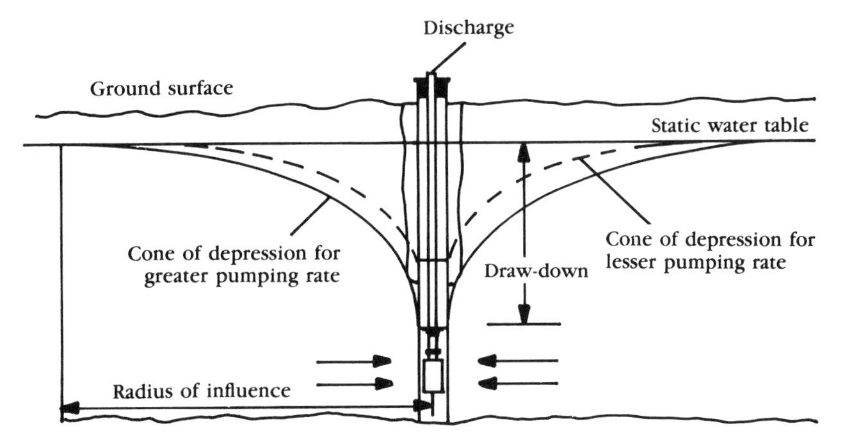

8-3 When water is drawn from a well, the removal creates a temporary cone of depression.

The shape of the cone depends upon the nature of the aquifer, whether it is highly or only slightly permeable, and the rate of water withdrawal from that particular well. Draw water too rapidly from the well and it might go dry or show reduced flow even though there is still plenty of water nearby. When two or more wells draw from the same aquifer, the maximum rate of flow into each will be reduced. The cone of depression formed around one well will extend to the adjoining well and so reduce the quantity of water available there.

SPRINGS AND ARTESIAN WELLS

Springs are formed when the lower end of an aquifer extends through the surface of the earth. Very often the aquifer extends through the side of a hill. From there, the aquifer is beneath the zone of aeration angling up toward the top of the hill. Rainwater penetrates the zone of aeration and seeps downward, escaping through the exposed end of the aquifer. Sometimes the aquifer is a formation of unconsolidated rock and the water seeps from exposed cracks. Depending upon the size and nature of the aquifer, the spring might be an all-year dependable source of good water or it might run only following a rain.

Artesian wells are named after the ancient town of Artois in France, which was formerly the old Roman city of Artesium. During the Middle Ages it was there that the best-flowing, or at least best-known, artesian wells were to be found.

Artesian wells differ from springs in the pressure that forces the water from the ground. In an ordinary spring, the water merely flows out of the aquifer by gravity. To be classified as artesian, the water must emerge under pressure. The pressure might be low, in which case it merely bubbles out, or it might actually gush up into the air under substantial pressure.

The mechanism is easy to understand. The flow of water from an artesian well is similar to a squirt of water emerging from a leak in a garden hose filled with water under pressure. In the case of an artesian well, the aquifer reaches high up a hill or mountainside. Where the well is drilled, the aquifer is covered by a layer of impermeable rock or clay. The water in the aquifer is under pressure. When a hole is drilled through the impermeable layer, the pressure drives the water up the bore and into the air. Deep wells are sometimes called artesian, but unless the water in them comes up of its own accord they are not true artesian wells.

Chapter 9

Water quantity and quality

There is no certain way to forecast the quantity or the quality of water that can be drawn from a well before it is dug or drilled. If you calculate how much water you need and learn what quantity of water can be drawn from neighboring wells, you can estimate your chances of meeting your needs. Water quality is likely to be similar as well. If your neighbor's flow rate is far less than you will require, drilling deeper might help. Then again, you might have to install a second well. In any case, it is important to know your requirements and the chances of filling them before you drill.

RATE OF CONSUMPTION

The common figure for residential water consumption is usually given as 50 to 60 gallons per day per person. The EPA (U.S. Environmental Protection Agency) recognizes several categories: single-family dwellings, 50–75 gallons; multiple-family apartments, 40 gallons; luxury dwellings, 100–150 gallons; and estates, 100–150 gallons per resident. The latter two figures may be a bit high, the former two too low, and various authorities differ in their estimates. However, with water-using devices like automatic dishwashers and clothes washers now found in nearly every home and apartment, the estimates shown in TABLE 9-1 seem reasonable and workable.

The estimated figure of 95 gallons of water per day per individual in a modern household does not include water used to wash the car, hose down the walks and driveway, or fill the pool. And, of course, the figure would be multiplied by a fractional factor to include the number of adults usually present in the household. When guests are in residence for any appreciable time, the figure would have to be increased accordingly.

9-1. Water Consumption Rate

Consumer	*Quantity per 24 hours (gallons)*
Human	10
Cow	7 to 15
Horse	5 to 10
Hog	12
Sheep	1
100 chickens	4
Lavatory	1.5 per use
Bathtub	30 per use
Toilet (tank type)	6 to 8 per use
Shower bath	6 to 8 per use
Dishwasher	25 to 35 per use
Clothes washer	30 to 40 per use

Thus:

1 person per day uses	10 gallons cooking, etc.
	6 gallons in the lavatory
	30 gallons in the tub
	24 gallons using the toilet
	15 gallons using the dishwasher once every 2 days
	10 gallons using the clothes washer once every 4 days
	95 gallons per day

There are a number of ways to assess lawn, garden, and farming water requirements, which must be added over and above the 95-gallon figure. Tables are available listing requirements for animals and poultry of all sorts, and may be keyed to local conditions. The quantity needed for lawn and garden watering depends upon size, climate, soil conditions and the kind of watering equipment used. A so-called drip irrigation system, for example, uses far less water than a big sprinkler head.

The common figure for residential lawn and garden water requirements is 2 inches of water per week. This might appear to be a great deal of water, but on average you need 120 gallons of water to produce one dry, edible pound of plant. In other words, a pound of dry beans costs you 120 gallons of water.

Two inches of water works out to 1.5 cubic feet of water on every square yard of surface. One cubic foot of water equals 7.48 gallons. Thus, you need 11.2 gallons of water for every square yard of your lawn or garden area. But don't forget rain. You can learn the average rainfall in inches to expect from local records. County extension services and other agricultural entities can also provide specific information, which, along with the known area of your lawn and garden, you can use to calculate your "farming" water needs. To keep accurate tabs on rainfall, you can use a commercial rain gauge, or place a straight-sided, open-topped can in an open but out-of-the-way spot on your property. Measure the depth of the water after every rainfall.

Another approach for lawns is to use the commonly accepted figure of ½ inch per hour per square (100 square feet) of surface area. This equals a rate of 30 gallons per square per hour. Thus, a lawn of 5000 square feet would be watered at a rate of 30 times 50 or 1500 gallons per hour. The time periods would vary, of course, depending upon local conditions.

In some locales potable treated water can be used only for domestic purposes—normal in-house purposes like cooking and bathing. Watering or irrigating from a municipal supply is not permitted.

Fire protection

Where individual water systems are installed, effective firefighting depends upon the facilities provided by the property owner. This is a large subject unto itself, not to be considered here. Insofar as well capacity is concerned, however, if it is to be at all useful for firefighting it should demonstrate by pumping test a capability of producing 8 to 10 gallons per minute continuously for at least 2 hours during the driest time of the year. A typical small individual pressurized water system will pump only about 200 or so gallons per hour, or about 3¼ to 3½ gallons per minute, sufficient only for knocking down a small fire just starting or wetting down adjacent buildings. Thus, pumping capacity must also be in the 8- to 10-gallon per minute range.

Providing the needed quantity

Let's assume there are four adults in your family and you rarely have houseguests. On the basis of the figure cited in TABLE 9-1, you will require 4 × 95 gallons of water per day, a total of 380 gallons per day. This works down to an hourly rate of 15.8 or about 16 gallons per hour that your well must supply.

The chances that you and your family will actually use 16 gallons per hour are slim. There is always the possibility that someone will be filling the bathtub while someone else is washing clothes and someone else will be drawing water for washing or whatever, but this would occur only occasionally and for only brief periods. Water consumption will never be evenly spaced over a 24-hour period, although there will be times when several faucets or valves will open more or less simultaneously. Whenever this happens, the well must supply the total immediate demand or water will flow at reduced volume into the various plumbing fixtures and appliances.

Referring to TABLE 9-2, you can see that the optimum required flow rate for a bathroom tub, a dishwasher, and a clothes washer totals 16 gpm (gallons per minute). If they all go into use at once, there is no water left (at proper pressure) for the remaining faucets and fixtures. You cannot solve the problem at this point, but you can approximate an answer.

9-2. Recommended Flow Rates and Minimum Pressures for Common Household Fixtures

Type of fixture	*Minimum required pressure*	*Desired flow rate*
Lavatory faucet	8 psi	3 gpm
⅜ sink faucet	10	4.5
½ sink faucet	5	4.5
Bathtub faucet	5	6
½ laundry tub	5	5
Dishwasher	8	5
Clothes washer	8	5
Shower	10	3
Toilet tank	10	5
Toilet flush valve	15	15–40
Sill cock plus 50′ garden hose	30	5

Start by estimating the probable flow rate from the well you are planning to dig or drill. Do this by learning the constant available flow rate produced by a nearby well. We will assume that your well will reach the same aquifer and that the two wells will not be so close together that the cone of depression around one will adversely affect the other. Or technically, that their *radii of influence* will not overlap.

Assume now that half the faucets and valves in your house might be turned on at the same moment. Again taking figures from TABLE 9-2, you come up with a total draw of, say, 28.5 gpm, which is just an educated guess at this point. If you estimate that your well will produce 30 gpm, then in theory it will never run dry. You require no more than a minimum-size storage tank, and that only to be on the safe side.

On the other hand, if you estimate your projected well will only produce 20 gpm, there almost certainly will be moments when the demand exceeds the supply. As stated in this example, there will be times when you will need water at a rate of 28.5 gallons per minute and you will need it long enough to fill the bathtub, dishwasher, clothes washer, toilet, and other figures or appliances. All in all, by the table, you are going to require more than 100 gallons in less than 5 or 6 minutes.

That amount is a little high because the flow of new well water into the system during that 5 or 6 minutes is not included. Also, appliances and fixtures do not take all their water at a short gulp, but over a time period often considerably longer than 5 or 6 minutes. In any case, this is the way you can estimate your water requirements, your water tank size, and just how adequately your proposed water well may fulfill your needs. Figure 9-1 shows how projected water use can be charted.

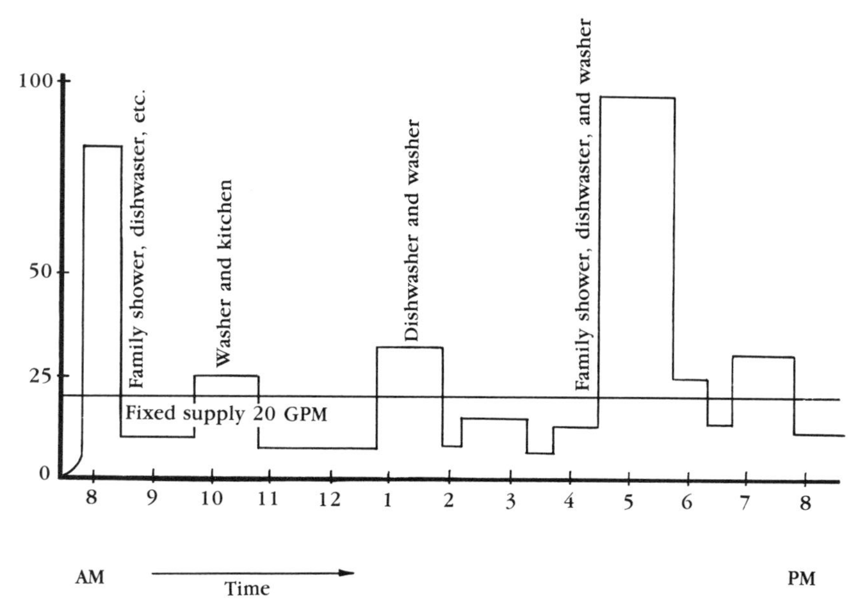

9-1 A graph of projected household water consumption will help you determine required water storage capacity and average need.

WATER PURITY

From our point of view it is important to evaluate water as having two separate characteristics: purity and quality. Hydrologists generally consider purity a function of water quality.

We define pure water as water that is fit and safe to drink—it is *potable*. Water might be crystal clear, cold, and even free-running, but still contain pathogens, microbes, and bacteria injurious to one's health. Generally, if your well is 50 to 100 feet away from the nearest source of contaminants, chances are fairly good that the water is safe to drink. The earth is a tremendous filter and purifier. However, this is never, ever a sure thing. To be safe, be certain: have your water supply tested, preferably once a year or so.

In most areas the local health or sanitation department stands ready and willing to check your water supply at little or no cost. Either an inspector will visit your home site and take samples, inspect the land for the presence of insecticides, the location of your proposed well in relation to the septic tank, barn, and the like, or the office will provide you with a sterile bottle and instructions on how to fill it (assuming the well is in) with a test sample of water.

They will make a bacterial and coliform count, which indicates the quantity of human and other warm-blooded animal feces present in the water. They will then inform you if the sample passes their standards, and if it doesn't, what steps you can take to correct the problem.

WATER QUALITY

Water does not remain in one place very long. It is in constant circulation from the land to the sea and back to the land again (FIG. 9-2). This constant movement is called the great water cycle, or *hydrologic cycle.*

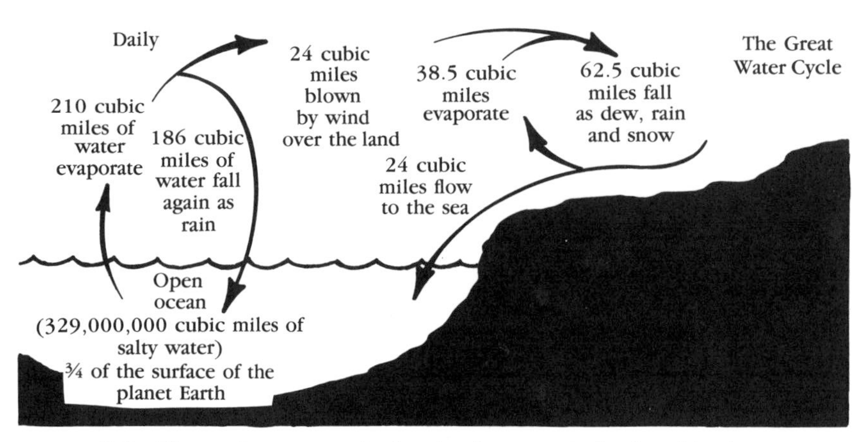

9-2 The cyclic movement of water from sea to land and back.

The oceans, which cover three-fourths of the earth's surface, contain an estimated 329 million cubic miles of salty water. Each day some 210 cubic miles of water enter the atmosphere by means of evaporation. Of this, about 186 cubic miles of water fall back into the seas as rain. The wind blows the additional 24 cubic miles of water over the land where it falls as rain or snow. The total quantity of rain, dew, and snow that falls on land each day amounts to 62.5 cubic miles. The rest, 38.5 cubic miles, evaporates from the land. An additional 24 cubic miles is made up of rivers and streams that empty into the oceans.

Water has the greatest surface tension of all common liquids except mercury. Surface tension is the nature of a liquid to try to pull itself into a ball (a drop of rain) and to adhere to a surface. When water falls upon the earth, capillary action pulls the water into the soil and spreads it out horizontally. If it were not for this characteristic, a far greater percentage would run off the surface and directly back into the seas instead of finding its way into aquifers or nourishing plant life.

The strangest characteristic of water is that it is almost a universal solvent, but is not itself changed by the substances it dissolves. By one method or another, it can always be recovered and used again. This is an especially remarkable property because water can dissolve so many disparate substances. For example, nearly all of the known elements have been found in sea water, some as mere traces and others in relatively large quantities.

We all know that surface water differs considerably from one source to another. The water in Soap Lake in central Washington feels soapy. Milk River in Montana looks milky. Stinking Water Creek in Nebraska has a smell. Some lakes are sweet, some are salty, and some are acid.

This is all due to the difference in quantities and kinds of dissolved minerals.

Pure (distilled) water with no dissolved minerals actually has a poor flavor. The taste of the water you draw from your well depends upon the nature of the aquifer and the distance the water has traveled through it. When the well taps into several interbedded aquifers, the quality of the water will vary from time to time, depending upon the particular aquifer most of the water is flowing through at the time. The depth to which the well descends may also affect the quality of the water. In northwestern North Dakota, where the earth contains sand, shale, siltstone, clay, and lignite, a relatively shallow well will produce hard (heavily mineralized) water. But in the same location, a well that reaches much farther down into the earth will produce soft water.

Hard water versus soft water

Hard water denotes water that does not readily form lather with ordinary soap. Any water that has a relatively high percentage of dissolved minerals is classified as hard water. Sea water, for example, is hard water. In addition to making washing difficult, hard water poses many other problems. It tends to deposit minerals in pots and pipes. The rate of deposition depends upon the percentage of minerals present and the temperature of the surface with which the water comes in contact. Hard water will form a layer of calcium on the inside of the tea kettle and the water boiler, but little if any on the walls of cold-water pipes.

In hot-water pipes, boilers, and water heaters, even a thin layer of stone reduces heating efficiency. A thick layer—and it can grow to ½ inch or more—practically insulates the water in a boiler from the applied heat. It also fouls valves and builds up crusts on faucets. In many cases the only practical solution is to replace pipes and parts, although some items can be successfully treated, if the buildup is not too bad, with hydrochloric acid. One recourse is to install a water softening system, which requires little attention or maintenance and removes most if not all the minerals from the water supply at the point of entry into the building. But one must consider the cost of the water softening system and the annual or semiannual servicing required, and the introduction of a small quantity of salt (ordinary table salt) into the water. This may not be acceptable if household members are on a restricted-salt diet. With additional pipes, the kitchen can be supplied with unsoftened water, or the softening system can serve only the domestic hot water supply.

To determine the hardness or softness of your water supply (TABLE 9-3)—beyond trying to estimate it by the relative difficulty of working up a soap lather—you can check with your neighbors, ask the local water-softening company (they usually provide free testing), and take a sample to your local board of health for testing. They often do not charge, or charge only a nominal fee.

The concentration of dissolved minerals in surface water may vary with time and the seasons. For example, the saline content of the Great

9-3. Hardness Comparison

Parts per million	
0–60	Soft
61–120	Moderately hard
121–180	Hard
More than 180	Very hard

Salt Lake in Utah was about 15 percent during the 1870s when the water level in the lake was high, but about 28 percent in the early 1900s and again in the 1960s when the level was low.

Groundwater mineral concentration varies only slightly. The major causes of hardness in water are calcium and magnesium, and other minerals, such as iron, manganese, aluminum, barium, strontium, and free acid also contribute to the hardness of water. Hardness is usually measured by the quantity of calcium carbonate (limestone) or its equivalent that is left behind when a specific amount of water has evaporated.

Water with fewer than 60 ppm of dissolved minerals is rated soft and is suitable for all purposes without treatment. Water in the 61 to 120 ppm range is moderately hard. It can be used for most applications with little difficulty. Exceptions are high-pressure steam boilers and other applications involving high temperatures. When the hardness range is above 121 ppm, you usually need some sort of softener or a special soap. Water in the hardness range of 180 ppm and higher requires water softening treatment. If the minerals are not removed, they can quickly clog pipes, valves, strainers and cooking utensils. Hard water is also hard on human skin, especially hands and face, and it shortens fabric life.

On the other hand, hard water is desirable for irrigation because it is readily absorbed by the soil. Soft water tends to form surface puddles.

EFFECT OF DISSOLVED MINERALS

Table 9-4 provides examples of the mineral constituents of various sources of water. Listed are the types of minerals and their relative percentage in ppm.

Silica, S_1O_2

The mineral occurs in quartz, sand, feldspar, and other rocks and many minerals. Surface water generally contains less than 5 ppm of silica, but a few samples can be found that approach 50 ppm. Ground water usually contains more silica than surface water. Silica is objectionable in water because it speeds the formation of boiler scale and forms interfering clumps on the blades of steam turbines.

9-4. Typical Minerals Found in Water

	Analyses of water, in parts per million, except for pH and color			
Constituent or property	*River*[1]	*Well*[2]	*Canal*[3]	*Lake*[4]
Silica (SiO_2)	5.4	41.	6.6	11.
Iron (Fe)	.11	.04	.11	.10
Calcium (Ca)	9.6	50	83	2.9
Magnesium (Mg)	2.4	4.8	6.7	9.5
Sodium (Na)	4.2	10	12	8690
Potassium (K)	1.1	5.1	1.2	138
Carbonate (CO_3)	0	0	0	3010
Bicarbonate (HCO_3)	26	172	263	3600
Sulfate (SO_4)	12	8.0	5.4	10,500
Chloride (Cl)	5.0	5.0	20	668
Fluoride (F)	.1	.4	.2	—
Nitrate (NO_3)	3.2	20	1.3	5.8
Dissolved solids	64	250	310	25,000
Hardness	34	145	235	46
Color	7	0	97	2
pH	6.9	7.9	7.7	9.8

[1]Stream in Connecticut
[2]Logan County, Colorado
[3]Drainage from the Everglades, Florida
[4]North-central North Dakota

Aluminum, Al

Auminum appears in quality only in water that has been in contact with bauxite and shales that contain high percentages of this metal. When the water contains a high organic content, aluminum will be found in concentrations of 1 ppm. Acidic water also can contain considerable aluminum in solution. Aluminum is only troublesome in the feed water of high-temperature boilers and steam turbines.

Iron, Fe

Iron is one of the three most common elements found in the top 50 miles of the earth's crust, and it is contained in many kinds of rock and soil (recognizable by the reddish colors). Surface water, unless it is acidic, rarely holds more than a few tenths of a ppm; if the water is acidic, it will often contain large quantities of dissolved iron. Groundwater, on the other hand, might contain only a few ppm of iron.

When the iron in nonacid water is exposed to the air, the iron oxidizes and settles out of solution. It takes very little iron to become visible. The concentration can be as low as 0.3 ppm and the iron will form red-brown stains on white china, plumbing fixtures, and clothing washed in the iron-containing water. It will also affect the taste of tea, coffee, mixed drinks, and other beverages. The recommended limit for iron content is 0.3 ppm.

Manganese, Mn

This element is another troublemaker. Like iron, concentrates of as little as 0.2 ppm can produce dark brown or black stains in fabrics and on laundry fixtures, and has a most unsettling effect on beverages. Manganese resembles iron in its chemical behavior. It is dissolved in appreciable quantities from rocks in certain sections of the country. Water impounded in dams will often remove the manganese from the mud at the bottom of the reservoir. Manganese is often found in water containing relatively high concentrations of iron. The recommended upper limit for manganese concentration in potable water is 0.05 ppm.

Calcium, Ca, and Magnesium, Mg

These two minerals are almost always found in water, because they are contained in almost all rock and soil. The highest concentration, however, results when water has been in contact with dolomite, limestone, and gypsum. These two minerals are mainly responsible for the formation of scale within boilers, water heaters, and associated piping.

Sodium, Na, and Potassium, K

Water that has not been distilled contains some quantity of sodium and potassium. When the concentration of these two elements is under 50 ppm, the overall usefulness of the water is not affected adversely. When the concentration goes above 100 ppm, the water will foam in a steam boiler. When the concentration is higher still and has a high proportion of sodium, the water should not be used for irrigation.

When the sodium content of household water must be pinpointed, a laboratory analysis should be made periodically. Ion-exchange water softeners are frequently used in supplies where the water is hard, increasing the amount of sodium. People on a low-sodium diet for various health reasons should be aware of this. A typical low-sodium allowance for drinking water is 20 ppm. If this limit is exceeded, changes in either the diet or the water supply arrangements may be required.

Carbonate, CO_3, and Bicarbonate, HCO_3

Groundwater rarely includes carbonate and bicarbonate. When water is treated with lime, some carbonate will be released and dissolved. Bicarbonate also results when water containing dissolved carbon dioxide makes contact with rocks. If the rock is granite or a similar composition, the concentration of bicarbonate rarely exceeds 25 ppm; at most it is never higher than 50 ppm. When water containing carbon dioxide gas contacts carbonate rocks the concentration of bicarbonate in the water can rise to 500 ppm. When the concentration of carbonate and bicarbonate is very high the water is classed as alkaline. This is usually expressed as ppm calcium carbonate.

Sulfate, SO_4

Very little sulfate, which takes many specific chemical forms, is generally found in rivers and wells. You find these minerals mainly where the water contacts beds of gypsum and shale. It is also found in acid-mine drainage and is formed by the oxidation of sulfides or iron. When present in large quantities, it produces a very hard scale within boilers. Water containing high levels of sulfate caused by the leaching of natural deposits of magnesium sulfate (Epsom salts) or sodium sulfate (Glauber's salt) can have undesirable laxative effects. The upper limit for sulfate concentration should be 250 ppm.

Chloride

Chloride can range as high as several hundred ppm in streams and lakes in semiarid regions and streams that carry irrigation runoff. Sewage also acts to increase the chloride content of water; a sudden rise in chloride level in a water supply may be an indicator of that problem. Most rock contains a percentage of chloride that can be dissolved by water. But chloride concentrations of less than 25 ppm usually present no problems. When the concentration is much higher, and the water also contains calcium and magnesium, the corrosive power of that water is greatly increased. Drinking water containing levels over 250 ppm of chloride usually has an unsavory flavor.

Fluoride, F

There is much argument, both factual and emotional and largely unresolved, about the efficacy of fluoride in drinking water. The concentration of fluoride found in rocks is about the same as for chloride, and there is very little if any in surface water. Where natural concentrations of optimum levels do occur, the incidence of dental caries has been found to be below the rate prevailing in areas without natural fluorides. That optimum level depends upon the average ambient temperature, because that is the primary influence on how much water people drink. Optimum concentrations are 0.7 to 1.2 ppm. Excessive concentrations may produce fluorosis (mottling) of teeth; the maximum recommended level is generally considered to be 2 ppm.

Nitrate, NO_3

Nitrate, any of numerous inorganic compounds derived from nitric acid, is considered the final oxidation of matter that contains nitrogen, and that includes all organic substances. Nitrate-containing water can be used for all industrial applications, but its use for human and animal consumption is dangerous. When the concentration of nitrogen in water approaches several ppm it is a strong indication that the water has been contaminated by sewage, livestock manure, fertilizers, or some other organic matter. Excess nitrate in drinking water is the cause of methemoglobinemia, or "blue baby disease." Water containing over 45 ppm of nitrate, or 10 ppm of nitrogen, should not be used as a domestic supply.

Nitrite, NO_2

Related to nitrate, this is a group of salts or esters of nitrous acid, such as potassium nitrite. It is sometimes present in polluted wells, and is a relatively sure indicator of pollution. Where concentrations exceed, or even approach, a level of 1 ppm, ingestion of the contaminated water can have very serious effects upon infants and should not be used.

Copper, Cu

Copper is found naturally in some water, especially where copper ore deposits have been mined. It may also be introduced into corrosive water that passes through copper pipes. In small amounts it is not considered detrimental to health, but it does impart a characteristic unpleasant taste to the water. The recommended limit is 1 ppm.

Zinc, Zn

Like copper, zinc is naturally found in some water, and is typically present in water where zinc mining has taken place. Serious ground water pollution problems have developed around both operating and abandoned mines, particularly those where silver, lead, and gold, all often associated with zinc, are or have been present; coal mines are also serious offenders. Because of weathering and leaching, heavy concentrations of iron, manganese, sulfates, acids, and other minerals have resulted.

Zinc is not considered detrimental to health, and in fact a certain amount is essential to good health. It does, however, impart a bad taste to drinking water and foods prepared with it, so the recommended limit is 5 ppm.

Lead, Pb

Lead may be found naturally in some water, typically by leaching from mine tailings or similar sources. More commonly, it is introduced into a water supply via lead pipes, soldered pipe connections, or contact with lead paints. Either a brief or a prolonged exposure to lead can lead to extremely serious health difficulties in humans; prolonged exposure can lead to death. Lead taken into the body in quantities in excess of certain relatively low normal limits is a cumulative poison. A maximum concentration of 0.05 ppm of lead in water must not be exceeded. Current thinking tends toward the idea that there is no truly safe level, and that drinking water must be lead-free.

Trace elements and toxicity

We do not fully understand the place of trace elements in human, animal, or plant existence. We do know that we need copper, cobalt, zinc, and other elements in our diets and that minute quantities, in terms of parts per billion, are necessary to health and even to life itself. On the other hand, they are intolerable in excessive amounts, the levels of which are really unknown, and the presence of other traces are poisonous to man. Some of the elements have already been discussed, such as fluoride and lead. But there are many more elements—beryllium, arsenic, and strontium, for example—and synthetically produced chemical compounds like fertilizers and pesticides that are known to be toxic, and still others that have certain undesirable characteristics that interfere with the use of water even when present in relatively small quantities. Because of widespread and ever-increasing appearance of these toxins in our water supplies, water quality has become a point of considerable concern and requires constant monitoring.

Turbidity

When water appears to be cloudy or a light cannot be directed through it without considerable loss, the water is turbid. This is caused by fine particles in suspension. The particles might be silt, clay, sand, or organic matter. Generally, when turbidity exceeds 5 percent the water should be filtered before use.

Taste and odor

Bad tastes and smells in water are usually caused by decaying organic matter and often indicate that the water is unsafe to drink. Sometimes a very high concentration of minerals—sulfur, for example—will also produce a bad taste or odor.

Dissolved solids

All the minerals that can be found dissolved in water are considered dissolved solids. Excluding the presence of a high concentration of toxic elements, water having no more than 500 ppm of dissolved solids can usually be used for all applications. Some special industrial processes, however, cannot tolerate dissolved solids exceeding 100 ppm. When the concentration exceeds 1000 ppm, the water is *saline*. When the concentration is higher than 35,000 ppm, the water is classified as *brine*.

pH

pH is a measure of the hydrogen ion content of the water, as well as a measure of the acid or alkaline content. Values are scaled from 0 to 14, 7 being neutral. Higher than 7 indicates increasing alkalinity, lower than 7, increasing acidity. Water in its natural state typically ranges from 5.5 to 9.0 pH. Knowing the value of a given water supply helps you select corrosion and disinfection control measures and determine proper chemical dosages.

ARTIFICIAL RECHARGING

When rain falls onto the earth, seeps through the zone of aeration, and enters an aquifer, the aquifer is recharged naturally. This, of course, is the usual way water enters an aquifer and reaches a well or spring. There are, however, times when it is advantageous to help nature recharge (see FIGS. 9-3 and 9-4).

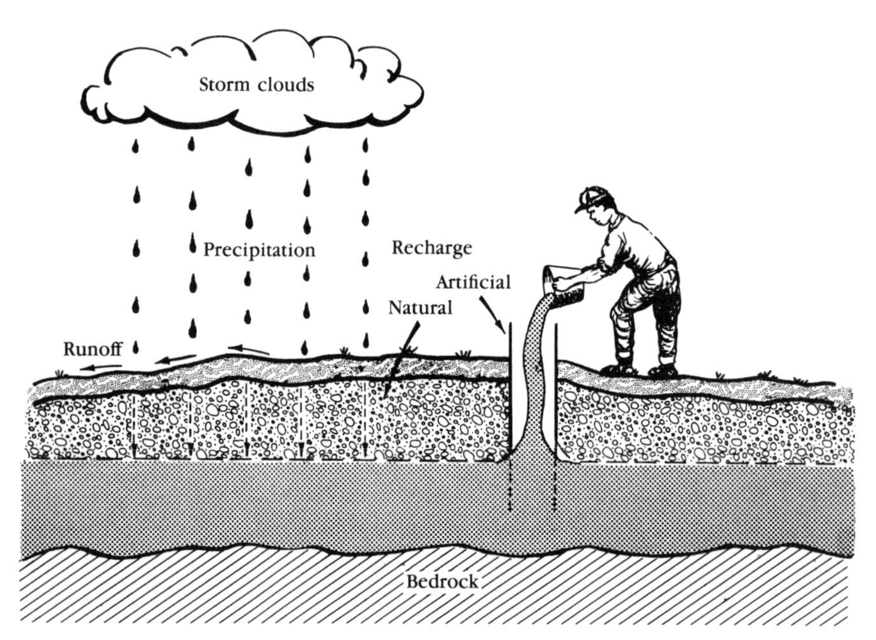

9-3 When man returns water to the aquifer, the process is termed artificial recharge.

Let's assume your well is going dry, but there is a free-flowing river nearby. The location and nature of the aquifer that feeds your well is insulated by clay or rock from the moving stream. You can recharge your well by drilling a second well 50 feet or so away from the first. Water from the creek is now pumped *into* the second well. Doing so recharges the first well. The reason for feeding the river water to a second well and

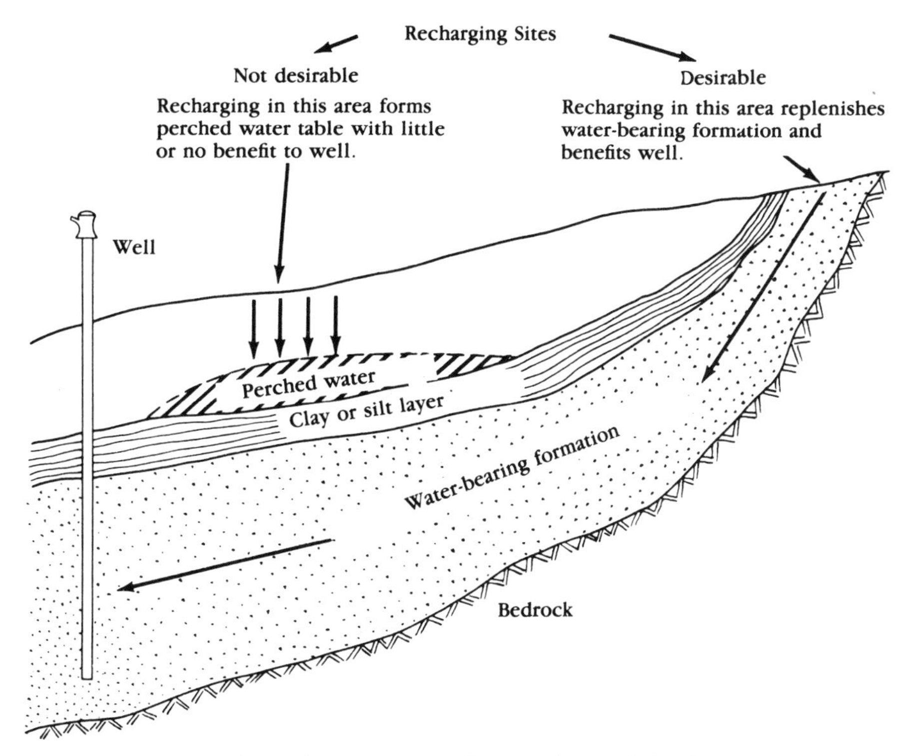

9-4 A site for recharging an aquifer must be carefully chosen.

not using it directly is that the river water needs treatment to make it potable. Flowing through the aquifer for 50 feet helps this process. Other reasons might be less cost and greater convenience in using the aquifer as a storage tank than constructing an unsightly above-ground or expensive underground tank.

The second well can hold much more water than the volume of its pipe alone. If, for example, you can draw 5 gpm from the first well, you can pour just about the same quantity into the second. The aquifer will soak it up.

Salt water intrusion

When a well is relatively close to a body of salt water and the well is pumped excessively, it is possible for salt water to find its way into the well. Fresh water is lighter than salt water. If the two are not mixed, they stratify; the lighter fresh water forms a layer on top of the salt water. As the fresh water is pumped out, the level of the salt water rises (FIG. 9-5).

When a well is located within reach of salt water, it is advisable to keep close tabs on the saline content of the water to forestall overpumping. In some instances, it may also be beneficial to artificially recharge the well to lower the level of salt water in the aquifer.

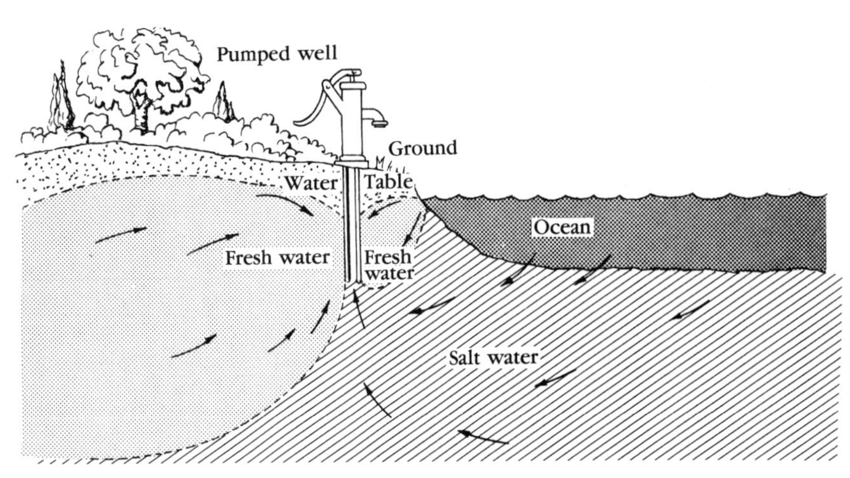

9-5 How salt water in a nearby lake or ocean can invade an aquifer.

For more information on the subject, write to the U. S. Department of Agriculture and request leaflet No. 452, *Replenishing Underground Water Supplies on the Farm.*

Chapter 10

Finding water

Approximately 90 percent of all the water to be found beneath the surface of the earth in this country is located within 200 feet of the surface. The average depth of residential water wells is about 50 feet. While these figures may be comforting, they do not assure you of finding water at a depth of 50 feet or, for that matter, finding it at all. In some areas, the water lies at tremendous depths—thousands of feet. How do you locate a site for a shallow well? How does one avoid drilling 200 feet or more to find usable quantities of water?

GEOLOGY AND GEOGRAPHY

The depth at which you will find an aquifer and the quality and quantity of water an aquifer can supply depends upon its location. The greater the total annual rainfall in the area, the greater the likelihood of water being found in copious quantities relatively near the surface of the earth. The western part of the United States is relatively arid, with an annual rainfall of less than 20 inches, while the eastern part is fairly dripping with rain, with an average annual rainfall of more than 30 inches, and is classified as humid. As FIG. 10-1 shows, the average annual rainfall in the United States varies from the desert, with less than 10 inches, to heavy rain areas with more than 60 inches per year.

Rain alone does not make a successful water well. The geology of the soil and subsoil makes a difference: all heavy rainfall areas do not lie above good aquifers. It takes a combination of rain and a suitable underground layer of sand, gravel, or unconsolidated rock to provide a well with water. Together, geology and geography are primary indicators of the chances of striking water.

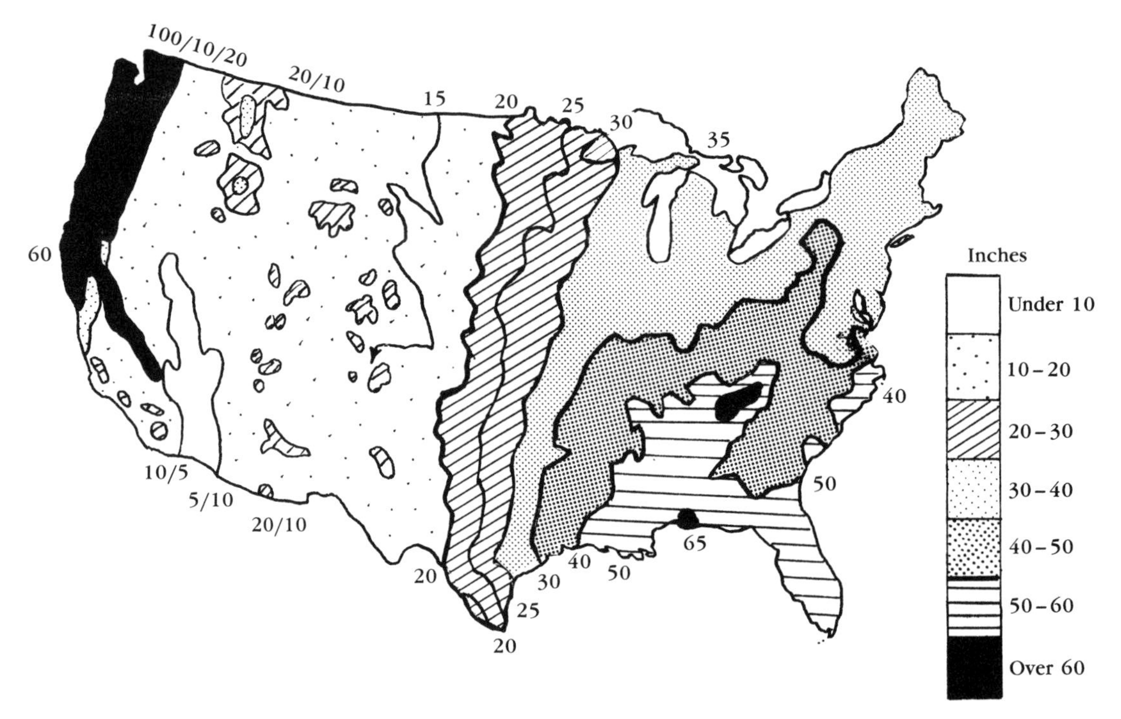

10-1 Average rainfall in the United States, showing areas capable of yielding 50 gpm or more from aquifers.

Relative elevations

Groundwater tends to follow the earth contours where the surface runs downhill. Your chances of finding water by digging your well at the top of a hill are far less than by digging at the bottom. But this is not invariably the case, because underground water can travel for hundreds of miles.

Topographic maps can be useful to help locate wells. A topographic map indicates relative elevations of the land and includes rivers, streams, lakes, marshes, and the like. Such a map will quickly tell you how high your property stands in relation to the rest of the world and where the nearby surface water is. Bookstores and sporting goods stores frequently carry maps covering the local area, and they can also be obtained direct from the U.S. Department of the Interior, Geological Survey, 1200 South Eads St., Arlington, VA 22202.

It is also a good idea to check with local, county, or state departments concerned with health or geology. In some areas, water maps have been drawn up that show local water supplies and conditions, and geologic maps that show the underlying strata of the earth can be a help as well.

Local water indicators

In addition to elevation and proximity to large bodies of water, other guides to finding suitable areas for a well include water seepage—continually wet spots indicating springs and seeps. You can drill your well into the same aquifer, positioning the well on ground higher than the seep. Also examine the bedrock that protrudes from the soil. Hard rock such as granite indicates very little chance of an aquifer. Your well has to intersect cracks in the rock to find water. The presence of soft stone like limestone or sandstone indicates a good possibility of finding water at 50 feet or less.

Plant indicators

A group of plants called *phreatophytes* can exist only where their root systems can reach the water table. The word phreatophyte comes from two Greek words meaning "well plant." Not only do these plants indicate groundwater, but to a certain extent they can tell you how much water is present and at what depth.

A willow or a cottonwood tree growing in a normally dry region indicates considerable water within about 20 feet of the surface. A good-sized tree can transpire several hundred gallons of water a day, so you can be reasonably certain there is good water in generous measure present in the earth when you see these trees. Some species of birch, sycamore, bay, live oak, alder, and red oak also indicate groundwater at a fairly shallow depth.

These trees lose their value as indicators in humid regions. There might be so much water in the zone of aeration that the plants do not have to reach deep into the earth to survive. Each region of our country has its own group of phreatophytes. Ask your local farm advisor, extension service, or agricultural experimentation station for their names. You can learn about them and study the characteristics and appearance of specific species in any of numerous wildflower, shrub, and tree field guides, readily available in most libraries and bookstores. Some of the more common phreatophytes found in arid areas of this country include:

- sedges, cattails, and rushes—good water near surface
- cane and reeds—good water 10 feet or less below surface
- saltbush—grows near salt water; check water quality
- pickleweed—mineralized surface water with 1–2 percent salt content
- arrow weed—good water 10–20 feet below surface
- small trees and elderberry shrubs—water no deeper than 10 feet
- rabbit bush—water above 15 feet
- black greasewood—mineralized water 10–40 feet down
- mesquite—good water 1–50 feet below surface

DOWSING FOR WATER

Some people claim to have been born with the ability to find underground water with the aid of a divining rod. The rod is most often a forked twig (FIG. 10-2) cut from a peach, willow, hazel, or witch hazel tree or bush. But sometimes the rod is no more than a straight twig. Sometimes it is made of metal, or is a pendulum or plumb bob.

10-2 The old "forked stick" technique.

The usual technique of holding the twig has the diviner's hands on the two portions of the fork, with the butt end pointing away from the diviner, although some reverse the position. Those operating with a straight twig hold it lightly by the thin end, straight out and horizontal to the ground.

Upon passing over underground water, the end of the forked twig will be attracted downward. Some diviners state that the twig will actually spring around in their hands, or that the pull is so strong that holding the twig straight is impossible. The distance the twig bends or dips and the strength of its response indicates the quantity and the depth of the water to be found directly beneath the divining rod.

You can, if you wish (when the neighbors aren't watching), cut yourself a forked twig, hold it gently in your hands (FIG. 10-3), and walk about your property until you feel a response. The plants useful for divining rods all require a great deal of water for growth and their roots will even grow sideways if necessary to reach it. The theory is that the twigs from those species have such a great affinity for water that they are naturally attracted to it.

Should the twig fail to respond in your hands, diviners will tell you that you have not been born with the gift of divination, or you are not

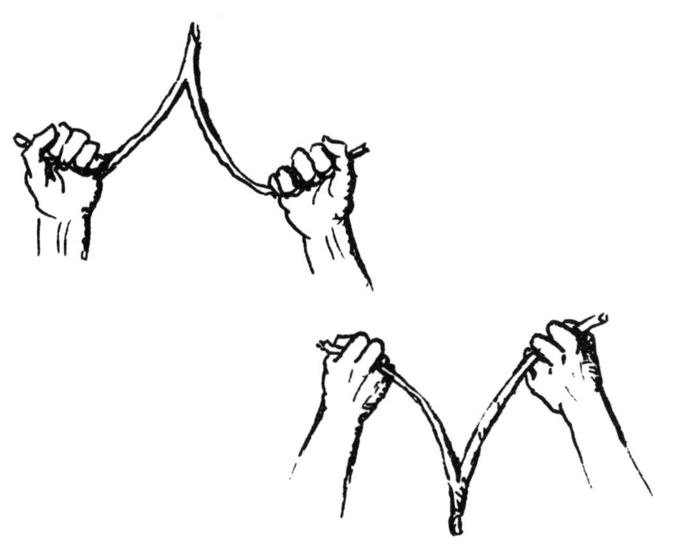

10-3 Two methods of holding a forked divining stick.

holding the twig properly. Holding it too loosely, they say, is as bad as holding it too tightly.

Should the twig respond vigorously above one spot on your property, have you truly found water? Should you dig or drill there? Because there is water to be found at some depth below the surface of almost the entire continental United States, and because considerable water exists at depths of 50 feet or so in much of this country, chances are fair that you have found water. On the other hand, the United States Geological Survey states categorically that "water witching," as dowsing is often called, is practically useless. They state that a tremendous amount of investigation has been bestowed upon the subject with an absolute lack of positive results. True enough, but they'll never convince a dowser.

The origin of the divining rod has been lost in history. Ancient writings indicate that "magic" wands of one kind or another were used by seers to find lost objects, forecast events, and find water. The divining rod is mentioned several times in the Bible, tying it in with wondrous performances, particularly in the Book of Moses. The Persians, Scythians, and the Medes used them. Marco Polo encountered the use of rods for divination in China. The ancient Germans, according to Tacitus, an early Roman historian, favored branches from fruit trees for dowsing. For more information on the interesting subject of dowsing, consult some of the many books and papers that are obtainable from libraries, or contact the American Society of Dowsers, Danville, VT 05828-0024.

MODERN TECHNIQUES

Professional well diggers use electrical resistivity measurement and seismic refraction (studies of induced vibrations in the earth) to help them determine the nature of the subsoil. They also use gamma-ray logging; this technique measures the natural radiation in a borehole.

OTHER SOURCES OF INFORMATION

If wells operate within a quarter or even a half mile, depending upon the local topography, discussing them with their owners could be time well spent. Learn how deep the wells are, how much water they yield, the water quality, the nature of the aquifer, and anything else that might be of interest. If there are a number of wells and you can obtain information on them, you can draw a sketch of the probable shape, depth, and extent of the aquifer, or at least the local portion of it. While it is possible to drill and miss an aquifer altogether, they are usually at least hundreds of yards wide. If you drill between two producing wells, chances are you will hit water at about the same depth.

If you are planning to dig or drill your own well, the local drillers might be unwilling to impart their information to you. They might give you their data for a fee. In any case, well drillers keep logs of well depth, the nature of the aquifer, the level where it occurs, the type and diameter of the well drilled, and the yield obtained. This is valuable information and well worth buying.

Commercial well drillers will also drill a test well for a fee. It might be worth your money to invest in a test bore to learn beforehand what you will encounter in the way of rock and the depth you will have to go to find water. Knowing what problems lie ahead might save you from bringing in the wrong equipment or wasting time with methods unsuited to the subsoil you have to penetrate. Table 10-1 shows some of the most advantageous water well combinations of type, size, depth, and formations.

10-1. Practical Depths, Usual Diameters, and Geologic Formations Suitable to Different Types of Wells

Type of well	*Depth*	*Diameter*	*Geologic formation*
Dug	0 to 50 feet.	3 to 20 feet.	*Suitable:* Clay, silt, sand, gravel, cemented gravel, boulders, soft sandstone, and soft, fractured limestone *Unsuitable:* Dense igneous rock
Bored	0 to 100 feet.	2 to 30 inches.	*Suitable:* Clay, silt, sand, gravel, boulders less than well diameter, soft sandstone, and soft, fractured limestone *Unsuitable:* Dense igneous rock

10-1 Continued

Driven	0 to 50 feet.	1¼ to 2 inches.	*Suitable:* Clay, silt, sand, fine gravel, and sandstone in thin layers *Unsuitable:* Cemented gravel, boulders, limestone, and dense igneous rock
Drilled:			
Cable tool	0 to 1000 feet.	4 to 18 inches.	*Suitable:* Clay, silt, sand, gravel, cemented gravel, boulders (in firm bedding), sandstone, limestone, and dense igneous rock
Rotary	0 to 1000 feet.	4 to 24 inches.	*Suitable:* Clay, silt, sand, gravel, cemented gravel, boulders (difficult), sandstone, limestone, and dense igneous rock
Jetted	0 to 100 feet.	4 to 12 inches.	*Suitable:* Clay, silt, sand, ¼-inch pea gravel *Unsuitable:* Cemented gravel, boulders, sandstone, limestone and dense igneous rock

Chapter 11

Before you dig or drill

You have found the spot for your well. All the indicators point to groundwater at your chosen location. Don't start drilling or digging yet. To just go ahead believing that not finding water is all you risk would be a serious error. You must first ascertain that you have the legal right to construct a well, and have complied with any pertinent laws. (In fact, your purchase of the property should be contingent upon obtaining that right.) And you must make as certain as you can that the water you finally do draw is not contaminated beyond redemption and unfit for use or consumption. This is not always easy to do, but there are steps you can take to protect yourself before you dig or drill your well, and even before you purchase the property

WATER RIGHTS

Just because you own a piece of property doesn't mean you also own any water that might lie on or under it. In fact, you may have no right to it whatsoever and it may belong to someone else.

The right of an individual to the use of available water varies from state to state. There are also different classes of use in some areas: domestic, irrigation or agricultural, and industrial. Some water rights do stem from ownership of land under- or overlying a water source, or even bordering it. Others, however, must be acquired by law. The vital conditions can't always be fulfilled and the right acquired. There are three basic types of water rights to consider.

The first is *riparian*: rights are acquired together with title to the land bordering or overlying the source. This is a common situation in parts of the country where water and rainfall are relatively plentiful, such as New England. Land underlying a body of water may belong, however, to the state, as may the water within it; neither can be simply taken over by a landowner.

The second is *appropriative*: the rights to either surface or groundwater are acquired following specific legal procedures that vary from place to place.

The third is *prescriptive*: rights are acquired by diverting and putting to use, for a period specified by statute, water to which other parties may or may not have prior claims. The procedures to be followed are governed by the water laws of individual States.

Purchase of a piece of property may include, and the contract may specifically state, certain water rights or water allotment for irrigating purposes. Water use on a property from a surface or groundwater supply may be restricted to domestic use only. A property may have irrigation water available, but no suitable domestic supply, or vice versa. To dig or drill you might need a permit, which could involve one or more hearings and substantial elapsed time, as well as cost. Arbitrarily tapping into a water source, even when it is on or under your own property, can lead to a very serious situation in many areas of the country, especially in the arid West where water law is king.

When purchasing property, make that purchase contingent upon the availability of water rights appropriate for your particular purpose: no one will guarantee sufficient quantity. When there is any question regarding the right to use water, a property owner should always consult the proper local authorities and clearly establish those rights.

WATER CONTAMINATION

The first step you can take to protect your well is to place it as far away as possible from any obvious sources of contamination. Unfortunately there is no positive safe distance. Groundwater can travel a long way, and if contamination seeps in at some point it can be carried to your well (FIG. 11-1). This is a constant threat, and periodic testing is a good idea.

Further precautions consist of positioning the well, if at all possible, uphill from any potential sources of contamination. Also, place the well beyond the drip line of a building roof and away from dry wells.

Great danger can exist in distant and otherwise hidden sources of contamination. Most people have heard the horror story of the Love Canal in New York, and how the toxic wastes dumped there many years ago have leached through the earth to poison the surrounding area. But the Love Canal is not the only toxic waste dump. There are thousands of them, and the locations of many of them are known only to the dumpers.

For a long time people believed in the "magic" of earth filtration and the work of its microbes. While microbes and the earth do a remarkable purifying and filtering job, microbes are not effective against inorganic substances and the earth is a filter with a finite, and variable, capacity. Sooner or later it loses its ability to filter, assuming that poisonous chemicals have to travel some distance through the earth and do not move only through the wide spaces found in coarse sand and gravel. Some substances cannot be removed in this manner at all.

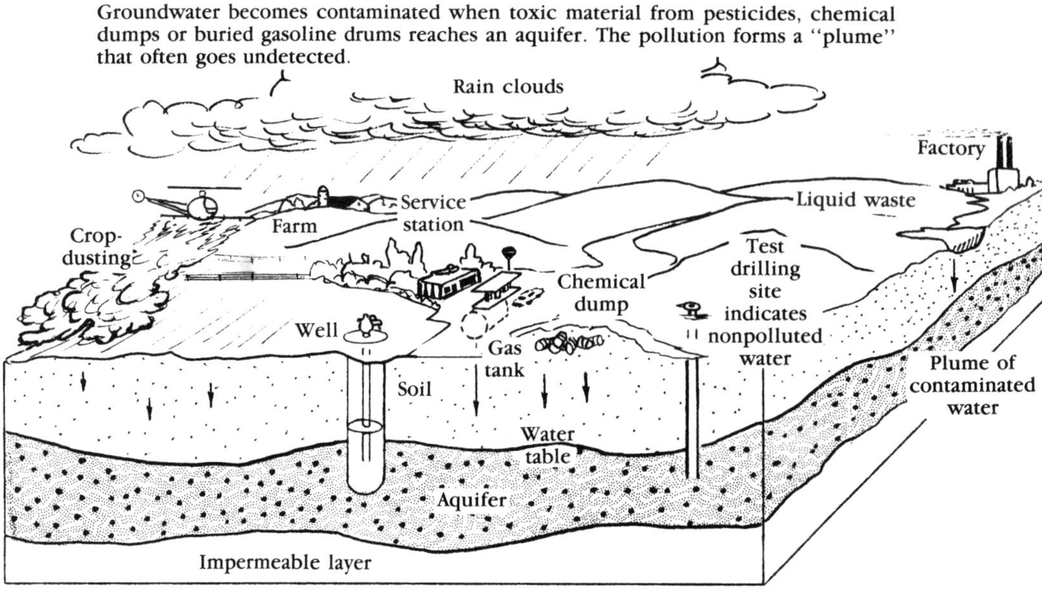

Groundwater becomes contaminated when toxic material from pesticides, chemical dumps or buried gasoline drums reaches an aquifer. The pollution forms a "plume" that often goes undetected.

11-1 How pollution can reach an aquifer.

There are more than 63,000 synthetic organic chemical compounds now being marketed, few of which an be "cleaned" by microbes. All are potentially lethal or at least harmful when ingested, even in minute quantities. In addition to the esoteric poisons, there are toxic lead, mercury, TCE (trichloroethane, a common dry cleaning agent and solvent), benzene, toluene, carcinogenic vinyl chloride, gasoline, carbon tetrachloride, aldicarb and DBCP (both pesticides).

There are at least 51,000 known toxic dumps in this country. Environmental Protection Agency officials estimate that three-quarters of abandoned and still-in-use dumps are leaking poisons into the ground. Thousands upon thousands of sanitary landfill operations—ordinary municipal trash and garbage dumps—are sources of heaven-knows-what toxic substances oozing into the groundwater. And once an aquifer is contaminated, there is no known practical way of cleaning it up. In some areas, very little is needed to contaminate an aquifer and make the drinking water dangerous. In Florida, where in some areas the water table is only 6 inches below the surface, runoff from a car wash can cause trouble.

The physical and mental difficulties produced by tainted water range from cancer to constant headaches, acute kidney problems, birth defects, nausea, skin rashes, brain tumors, and a whole medical dictionary of other ailments that can be merely annoying or terminal. Former EPA official Eckhardt Beck says, "The contamination of the groundwater is the environmental horror story of the '80s." Unfortunately, the story is continuing into the '90s, shows no sign of ending, and may be worsening as more dumps are discovered and illegal dumping continues.

Millions affected

Although no more than 2 percent of the U.S. population is presently affected by contaminated drinking water, the total number is easily 5 million. Throughout the nation, more than 2000 wells have been shut down. Federal budgetary constraints and bureaucratic red tape have hampered clean up and preventive measures, and local officials are not equipped to handle such problems. Local standards of purity for wells range are sometimes nonexistent. The number of suspected toxic dump sites far outstrips the investigation and inspection capabilities currently in place.

Poisoned water is difficult to detect. Without testing, you cannot tell bad water from good until it becomes so contaminated that you can taste the poison (and many contaminants are tasteless). Until that point the water might seem clear, cold, and sparkling clean.

Distance from a toxic dump is no guarantee that your well will not be affected. Toxic waste first dumped on a regular basis into a pit 90 miles outside of Memphis in 1964 reached the water system of that city in the late '70s. Citizens became ill from drinking their tap water and even passed out in the shower.

Look before you dig

The best way to protect yourself and your family against contaminated drinking water is to check the proposed well site carefully before you begin any well work. Check for possible sources of surface contamination: septic tanks, drain fields, barns or feed lots, and remember the minimum separation requirements (TABLE 11-1). Obvious sources of

11-1. Distances from Wells to Sources of Contamination

Formations	*Minimum acceptable distance from well to source of contamination*
Favorable (unconsolidated)	50 feet. Lesser distances only on health department approval following comprehensive sanitary survey of proposed site and immediate surroundings.
Unknown	50 feet only after comprehensive geological survey of the site and its surroundings has established, to the satisfaction of the health agency, that favorable formations do exist.
Poor (consolidated)	Safe distances can be established only following both the comprehensive geological and comprehensive sanitary surveys. These surveys also permit determining the direction in which a well may be located with respect to sources of contamination. In no case should the acceptable distance be less than 50 feet.

contamination include nearby gas stations or operations that have underground storage tanks for fuels or other liquids that might seep into the surrounding earth. Check with your local health department; they should know the current sanitary condition of the aquifers beneath your property. You can also request a complete sanitary survey.

If you have further doubts, go to the expense of having a test well drilled to sample the water. Have the water tested. This is neither simple nor cheap, and running a full range of tests might cost more than a thousand dollars, plus the drilling fee. If possible, try to pinpoint what the lab should look for by talking with neighbors, local residents, and the health department. This can cut costs down.

In any case, don't just guess at the quality of the water that comes up, or take someone else's word for it. Never depend on your sensitive palate or sense of smell. Poisoned water can be insidious: it often takes a long time to kill.

Chapter 12

Driven wells

When geological conditions are right, a driven well (FIG. 12-1) is the easiest, fastest, and simplest of all wells to construct. It consists simply of a pipe tipped with a combination well screen and point that can be driven into the ground with hand or power tools.

Points and pipes are typically 1¼- or 2-inch diameter, easy to handle, and can be driven 50 feet or more into the ground under good conditions. Although frequently considered to be useful primarily for vacation cabins, they can be equally suitable for year-round homes. Their cost is so low that two or more wells can be driven and tied together to achieve a more substantial output.

COMPONENTS

A driven well has only three main components: the point/screen assembly, the well pipe, and a pump. The supply line connecting the well to the pump is simply a continuation of the well pipe.

Well point

The well point (FIG. 12-2) is a specially constructed section of pipe with a reinforced steel or bronze tip at one end, a pipe thread at the other, and a screen of one sort or another in between. Points are selected with screen openings suitable to the aquifer being tapped. For coarse gravel, a screen with wide openings is chosen. For fine sand, a screen with very fine openings is required. There is always a compromise between reducing inflow of water and admitting sand (see FIG. 12-3 and TABLE 12-1).

When ordering a well point, you must specify the internal diameter or trade diameter of the pipe you intend to use with it. You will also need to specify the screen-opening size and describe the nature of the aquifer: gravel, fine sand, or coarse sand.

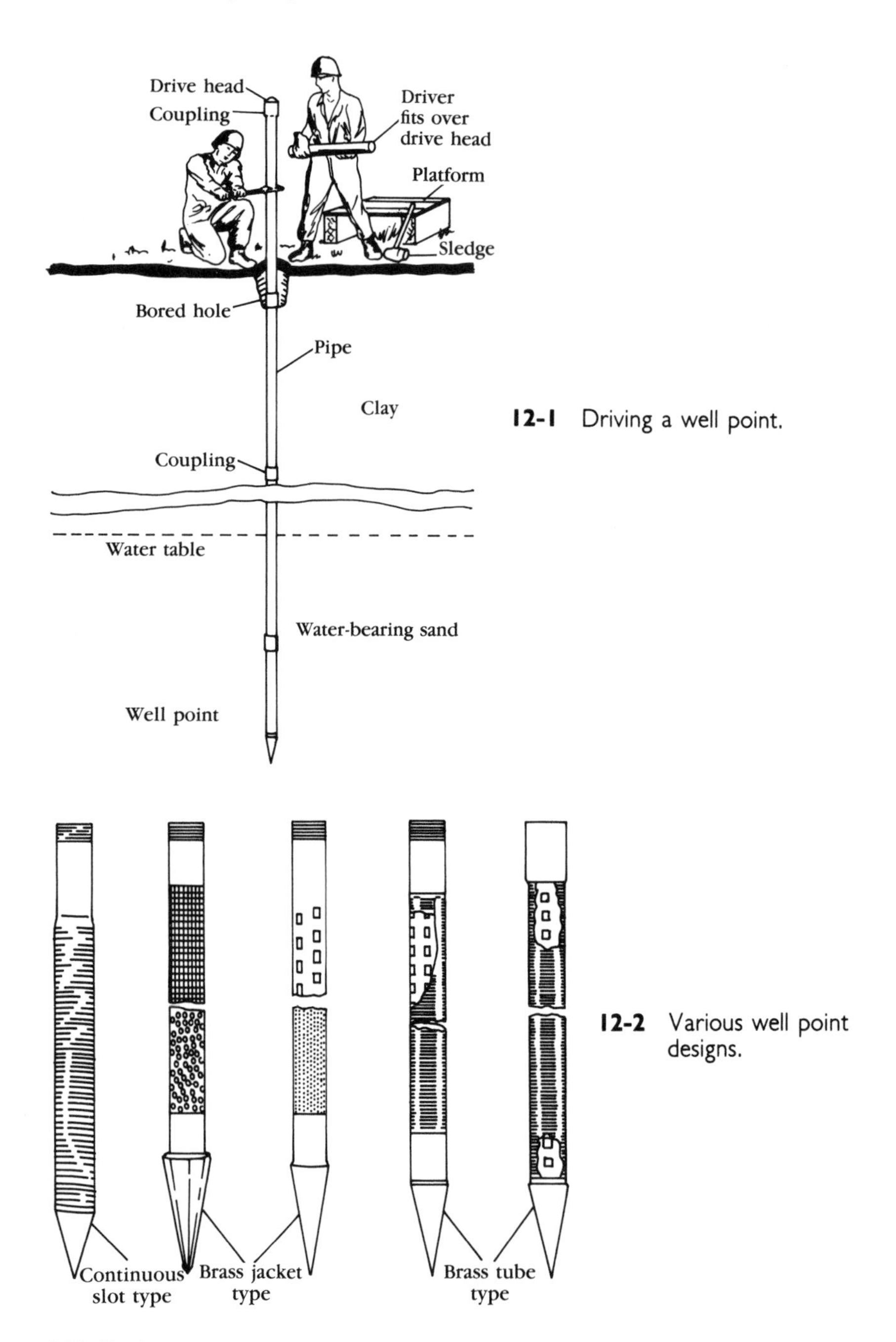

12-1 Driving a well point.

12-2 Various well point designs.

Well pipe

Two types of pipe can be used, depending upon conditions: standard galvanized water pipe or special well-point pipe called *risers*. You must use one or the other, because the two have incompatible threads and will not couple with one another (FIG. 12-4).

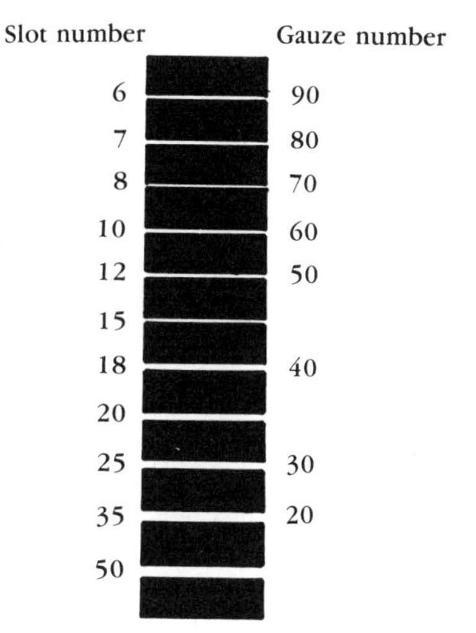

12-3 Standard slot numbers and equivalent gauze numbers. The white spaces represent the spaces to be found in a well screen with that number.

12-1. Steel-wound, Open-end Well Screens

Pipe length wrapping length (inches)	*I.D. pipe size (inches)*	*Pipe length wrapping length (inches)*	*I.D. pipe size (inches)*
30 × 24		60 × 54	
36 × 30		72 × 66	2½
48 × 42		96 × 90	
60 × 54	1¼	120 × 114	
72 × 66			
96 × 90		60 × 54	3
120 × 114		72 × 66	
		120 × 144	
48 × 42			
60 × 54	1½	48 × 36	
120 × 114		60 × 48	
		72 × 60	4
36 × 30		96 × 84	
48 × 42		120 × 108	
60 × 54	2		
72 × 66			
96 × 90			
120 × 114			

Courtesy Wesco, Inc.

Galvanized water pipe will work satisfactorily for shallow wells. However, it will rust through with time, usually at the threads, and it will not stand much heavy pounding during the driving. A total length of 20 feet or so is about the limit. Standard pipe comes in 10- or 20-foot

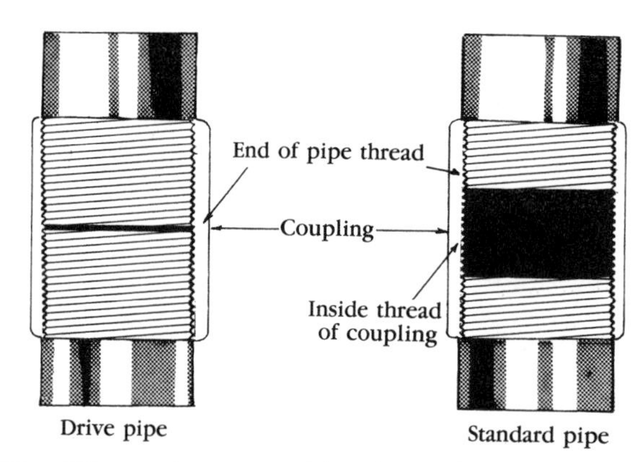

12-4 The difference between a drive-pipe and a standard pipe coupling.

lengths, much too long for driving. You must cut them into 4- or 5-foot sections and thread the ends and obtain extra couplings. Riser pipe, on the other hand, is manufactured in 5- and 6-foot lengths, and cutting is not usually required. When a short length is needed, it may be obtained ready-cut from the supplier. This is tough, stout pipe that will drive deep without failing, and is the choice for deeper driven wells.

Pump

Your choice of pump will depend upon the depth at which the water table exists and how far the water level will drop within the well pipe when pumping commences, whether or not electrical power is available at the site, and the diameter of the well pipe.

If the water level in the pipe will not drop below about 20 feet and the well is at or near sea level, a suction-type pump will be satisfactory. It can be motor-driven or a standard hand-operated pump like the traditional pitcher pump. Above sea level, the 20-foot operating depth will be reduced proportionally for vacuum or suction pumps. This figure might seem low in view of the ability of air pressure at sea level to support a 32-foot column of water, but no pump is perfect and air leaks hold the operating depth down.

When the water level in the well is or might be below 20 feet, you can install a special pump, either hand- or motor-operated, to overcome the lift difficulty.

WELL CONSTRUCTION

A driven well is simple, inexpensive, and can be built with a few hand tools and ordinary muscle power, provided conditions are right. In suitable soil, hand power alone can drive the well point down 30 feet or so. With power assistance, a depth of 50 feet or a bit more can be reached.

The one hitch is "provided conditions are right." If the well point encounters even a small stone head-on—say, 6 inches across—driving will stop abruptly. The point and pipe must be pulled. Then the stone must be removed if possible, or the well site shifted. If there are a lot of stones in the area, and especially if they are mixed with fine clay, driving becomes difficult, if not impossible. When too much resistance is encountered, the top of the pipe or the tip of the well point can be crushed or the pipe can buckle. The most suitable locations are areas containing alluvial deposits of high permeability.

Tools and equipment

You will need two pipe wrenches amply large enough to handle the pipe you will use, a small spirit level, pipe thread compound, and a length of fishing line with a lead weight at one end (to determine water level). You will also need some means to temporarily extract the water. Any small suction or jet pump, hand- or motor-operated, with a flexible hose that you can insert into the well, will do the job.

To drive the pipe you will need a drive cap (FIG. 12-5) that screws onto the top end of the riser or pipe. The cap takes the impact (and suffers the damage) from the driving tool and prevents damage to the pipe threads, which would make coupling lengths together impossible. If you drive by hand, a sledge or maul will work, or you can use a special impact driver, shown in FIG. 12-6. This steel tube fits loosely over the well pipe and drive cap. Several feet long, the upper half is solid and

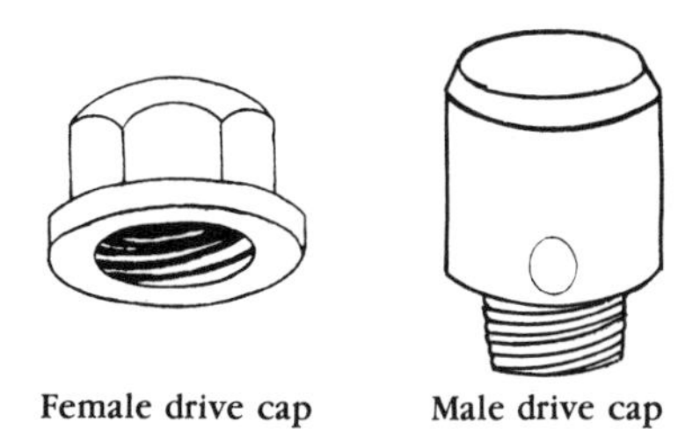

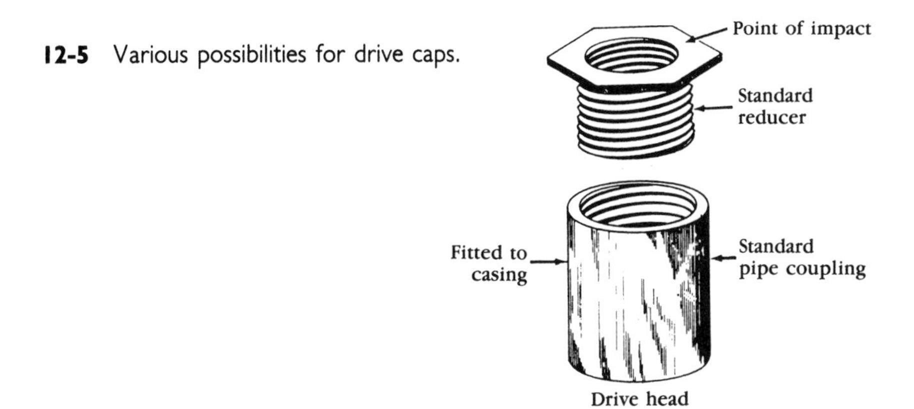

12-5 Various possibilities for drive caps.

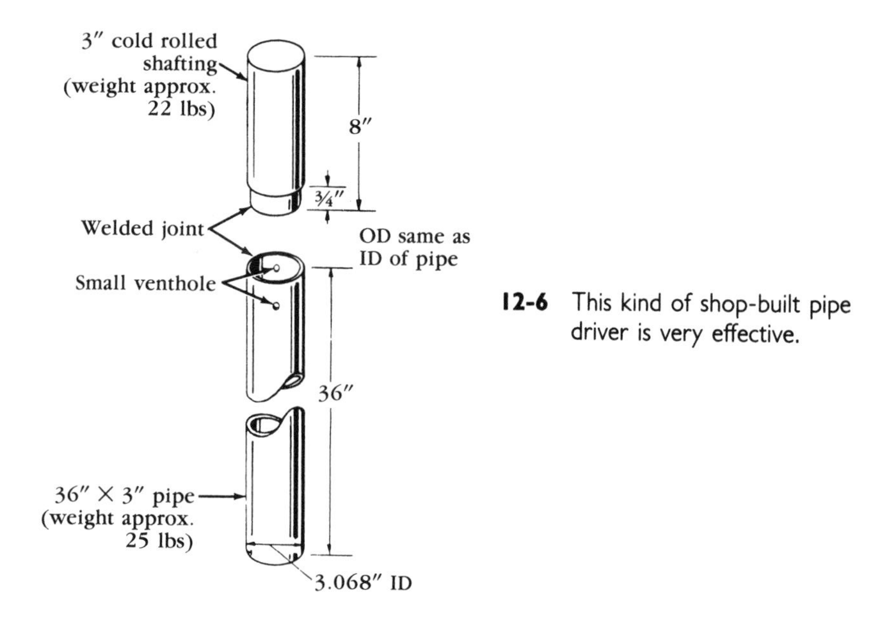

12-6 This kind of shop-built pipe driver is very effective.

heavy. It is operated by raising it as high as possible, then slamming it down on the drive cap. One advantage of this system is that you can't miss the cap, as you can with a sledge. The same is true of the internal impact driver shown in FIG. 12-7.

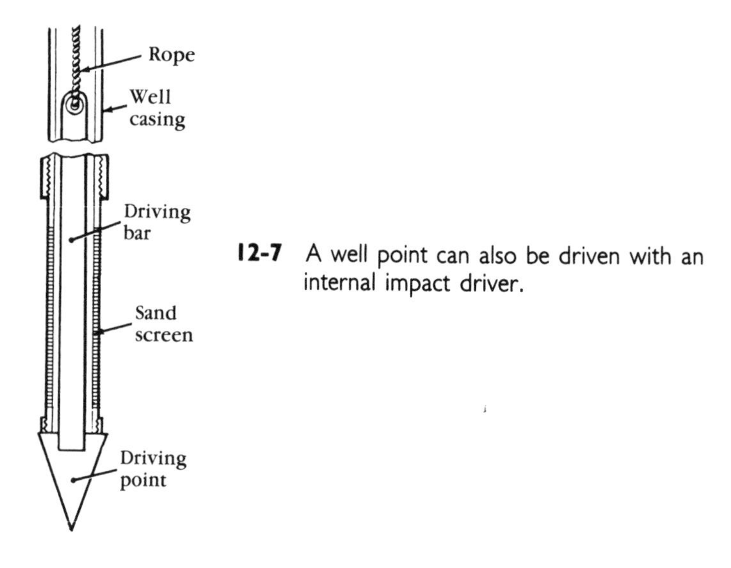

12-7 A well point can also be driven with an internal impact driver.

Using mechanical assistance makes driving easier. One method is to erect a tripod, A-frame, or derrick over the top of the well pipe, as in FIG. 12-8. A pulley or sheave is hung from the apex, with a rope run through the sheave. A heavy weight is fastened to the free rope end. Pulling the

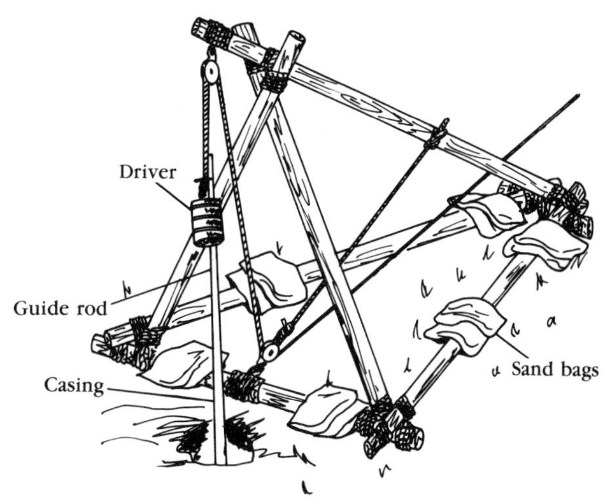

12-8 A low-tech tripod arrangement using only logs, rope, and sand-bag weights.

other end raises the weight. When released at the top of its travel, gravity takes over and the weight hammers down on the drive cap.

Multiple sheaves increase the mechanical advantage of the system; weights ranging from 30 to 300 pounds can be used easily. A wide weight will remain relatively well aligned with the drive cap. Smaller ones might need some sort of guide to keep them properly set. A guided weight striking a drive head (FIG. 12-9) is an even better arrangement. The drop-weight system is fully adjustable as to height and placement, and two or more workers can operate it together.

A still better arrangement removes most of the human labor while using almost the same set up. You can substitute a donkey engine of one sort or another to operate a smooth hoist drum to coil the rope and raise the weight. A typical drum is 8 to 14 inches in diameter and should operate at about 3 to 6 rpm (revolutions per minute).

Some farm tractors are fitted with such a drum for lifting hay into the barn and similar operations. A low-rpm drum can also be rigged up with a gas engine coupled to an old auto transmission, or a pair of transmissions coupled together, with the drum coupled to the output shaft. Or, an auto itself can be used. Jack one drive wheel clear of the ground and block the remaining wheels solidly. Bolt on a wheel rim (no tire), set the engine speed to the lowest possible idle rpm and put the transmission in reverse. Alternately tightening and releasing a rope looped once around the wheel rim does the job (FIG. 12-10); the trick is to release at the right time and not get the rope jammed or tangled. Commercial equipment designed for this sort of work has an automatic release mechanism on the drum drive. It can be set to rotate the drum just so far and then release. The drum speed can also be easily varied to suit conditions.

Starting the well is easier if you make a small starter hole to set the first pipe length into. You can do this with an auger somewhat larger in

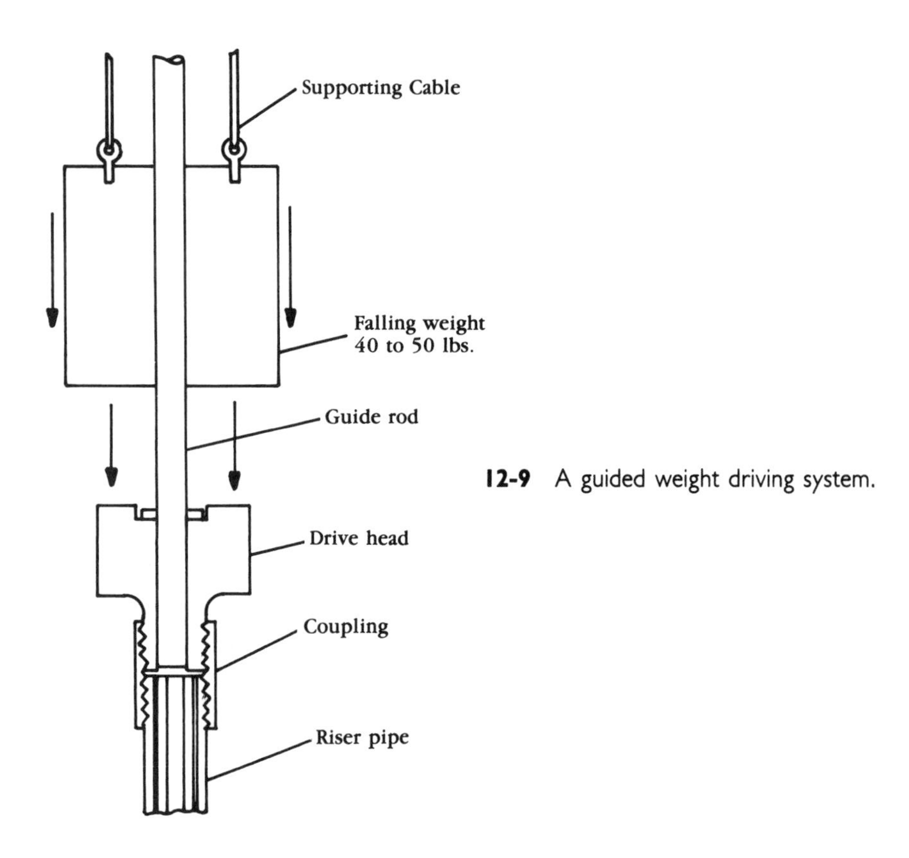

12-9 A guided weight driving system.

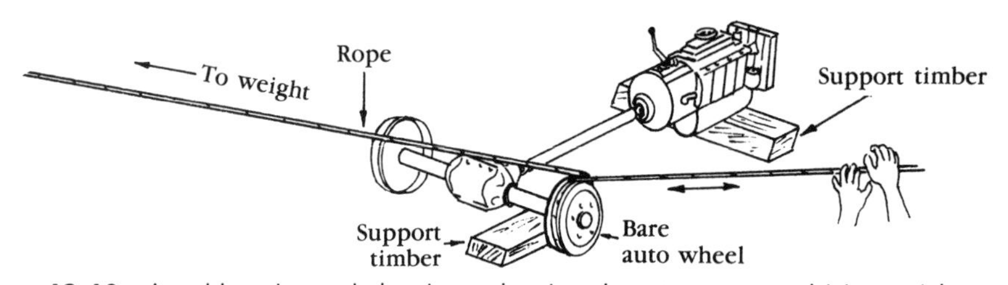

12-10 An old engine and chassis can be rigged up to power a driving weight.

diameter than your well pipe, boring down into the earth a foot or so. Or dig a hole a couple of feet deep with a posthole digger and a posthole bar if necessary.

To remove well pipe from the hole, you need at a minimum a sturdy pipe clamp that is strong enough to take upward strokes of a sledge and large enough to fit around your well pipe (FIG. 12-11). Or set a pair of vehicle jacks on wood blocks as shown in FIG. 12-12. They will easily lift the pipe free, but the clamp and jacks have to be frequently reset.

Another possibility is to set up a tripod over the well and hang a rope or chain fall to it. Do not attempt to use the fall and tripod alone to

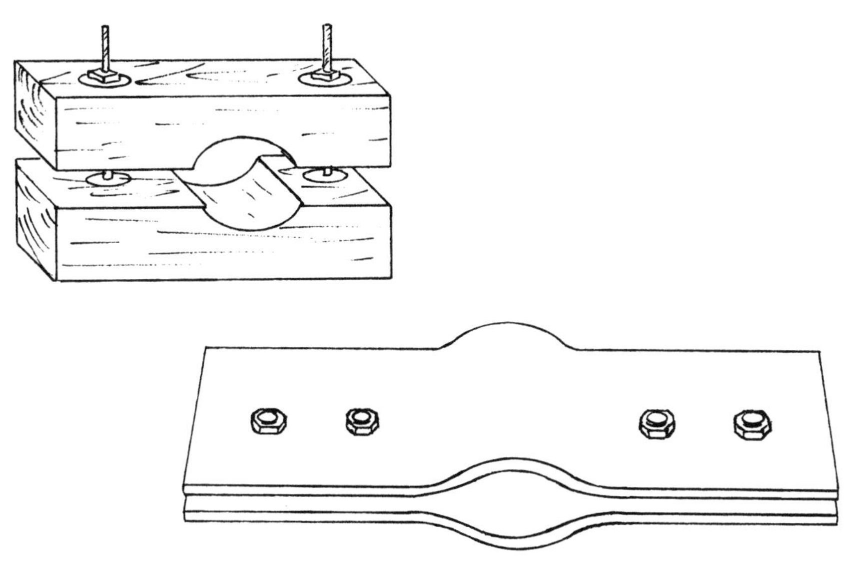

12-11 Two well-pipe clamps that can be fabricated from common materials.

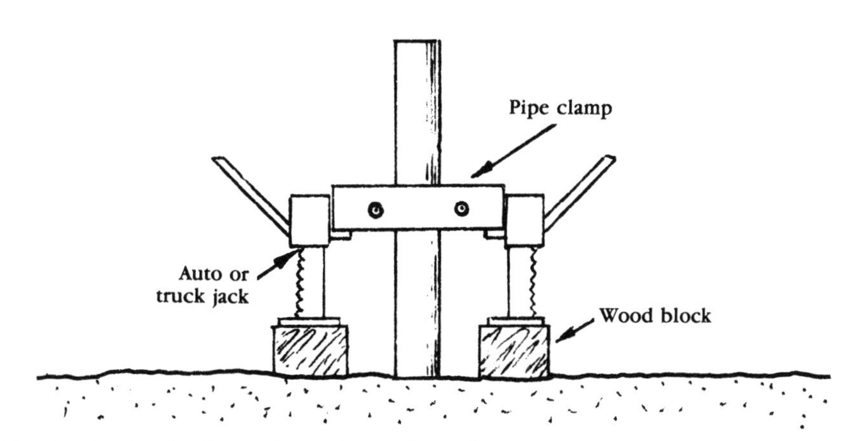

12-12 A pair of jacks can be used to pull a well pipe or casing by pushing it, in spite of the contradiction in terms.

pull the pipe. You can generate enormous pressure that could collapse the tripod or damage the pipe, or for that matter damage yourself. Use the set up only as an aid. Apply a little pressure with the fall, tap or jack upward a bit under the pipe clamp, and reset and repeat the process as necessary.

Driving the well

Start by boring a hole into the earth with a hand auger held as close to vertical as you can manage. A spirit level will help. If the auger cants, do not try to force it back into a vertical line. Back the bit out and start over. If you cannot get the first hole vertical, make a new one close by. The hole should be about ½ inch larger all around than the couplings of your well pipe. If an auger is not available, dig a hole about 2 feet deep with a narrow-bladed posthole digger, making the hole as straight up and down as you can.

Liberally coat the threads on one end of a length of pipe or riser with pipe joint compound and screw the well point onto it. The compound will make the threads run easier and, most importantly, make a water- and air tight joint. Use your pipe wrenches to tighten the joint as much as you can.

Cover the other end of this first section of pipe or riser with the drive cap; if threaded on, run the cap up hand-tight and then back off half a turn. Insert the well point into the auger hole and stand the pipe upright. Use your level to make sure it is plumb (vertical). A centering level, which is round with a flat bottom, works well for this. Place the level on top of the cap; when the bubble is centered in the circles, the pipe is plumb in all directions. Use dirt, wood wedges, or small stones to hold the pipe in alignment (FIG. 12-13).

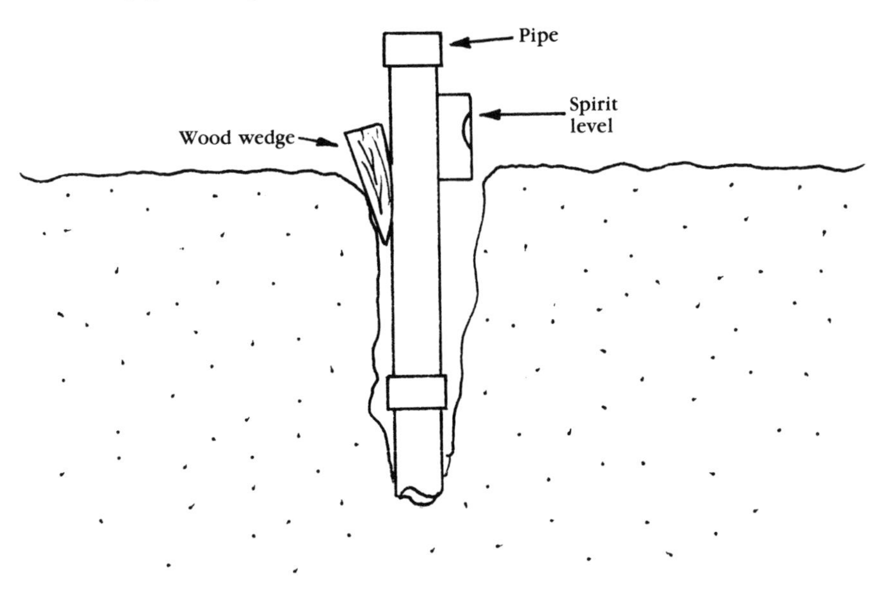

12-13 A spirit level and a wedge or two of wood can be used to plumb a well pipe or casing and hold it vertical.

With a sledge or maul, gently tap the point down into the earth, a bit at a time. At the outset you want to do no more than get the point properly started. Check constantly to make sure the pipe remains plumb. Getting this first section vertical is crucial. When the well point has disappeared from view, use a pipe wrench to turn the pipe gently (clockwise) to make sure the point is still tight. Do *not* rotate the pipe, as that could damage the point.

Once the first length of pipe or riser is firmly set in the earth, buried well and surely plumb, you can use more force to drive it. When the first pipe section is almost fully driven, gently turn the pipe again with a wrench to check the tightness of the point. Remove the drive cap, apply a coating of pipe joint compound to the threads, and run on a coupling. Dope the threads of a second length of pipe and screw it into the upper half of the coupling. Use both wrenches to take that joint up tight, but do not inadvertently rotate the pipeline in the process. Continue driving, and stop every once in a while to turn the pipeline just a bit to make sure the point and couplings remain tight.

If you drive with a block rather than a sledge, the weight you use depends upon the means you use to lift it, the diameter of well pipe you've selected, and the nature of the soil. If you will be working alone, choose a weight you can handle and lift over a period of time. Two people working together or in short shifts can use a heavier weight. If an engine will provide the power, the weight can be heavier still. But too much weight on a small-diameter pipe can buckle the pipe or mash the cap, or both. Prudence is advised.

The drop distance of the weight affects the drive cap as well as the pipe. The farther the block drops, the heavier the blow. Again, too much force can damage the pipe, or cause it to vibrate. If the weight drops again on a still vibrating pipe, the blow will not be centered and can bend the pipe.

A light or medium weight dropped too far on a stubborn pipeline may bounce back up into the air, and you will lose time by stopping to let the weight settle down. You are seeking a smooth rhythm; when you match the weight to the nature of the earth at the tip of the well point and adjust the drop distance of the weight and the drop-interval timing correctly, the weight will bounce cleanly upward just as you or the machine takes up the slack in the haul rope.

When you set up to drive, and periodically as you continue, use the block as a plumb bob. Let it come to rest and then adjust the tripod or derrick so when the weight is lowered directly onto the pipe drive cap it will balance there of its own accord. Centering the weight (which may not be the same as centering the block) is crucial if the pipe is to be driven straight and remain undamaged.

The driving rate refers to the distance into the earth achieved with each drop of the weight or swing of the sledge. In soft soil or sand, 2 to 3 inches per blow is about right. In "stiffer" soils the rate will be correspondingly less. In fine sand or clay the driving will be easier if you introduce water around the pipe.

Subsoil assessment

A knowledge of the characteristics of the soils your well point is penetrating is very useful. For example, it will help you determine when the well point has reached the maximum practical depth. The information will also help you select the best screen openings for the point. This may sound like advice to shut the barn door after the horse has gone, but it is not. In the absence of prior knowledge of the aquifer gained from other sources, you have no way to determine the character of the aquifer except by making a test bore or by making a judgment according to the details outlined in TABLE 12-2. To make these assessments, you must continuously monitor the sound the pipe does or doesn't make when struck; the depth of penetration with each blow; the distance of block rebound; and whether or not the pipe will rotate in the ground under relatively easy pressure from a pipe wrench.

If your driven well produces the flow of water you want there is no problem. If it does not produce sufficient water even after developing, the trouble might be undersized screen openings. On the other hand, going to the largest screen openings might fill your pump with sand. In addition to judging the subsoil characteristics by observing the action of the block, if necessary, you can withdraw the pipeline and well point and examine the subsoil stuck to the line. This can help you select a screen better suited to this aquifer.

However, before you decide to "pull" the point, try developing the well as described in Chapter 18.

12-2. Subsoil Drilling Characteristics

Type of formation	*Driving conditions*	*Rate of descent*	*Sound of blow*	*Rebound*	*Resistance to rotation*
Soft moist clay	Easy driving	Rapid	Dull	None	Slight but continuous
Tough hardened clay	Difficult driving	Slow but steady	None	Frequent rebounding	Considerable
Fine sand	Difficult driving	Varied	None	Frequent rebounding	Slight
Coarse sand	Easy driving (especially when saturated with water)	Unsteady irregular penetration for successive blows	Dull	None	Rotation is easy and accompanied by a gritty sound
Gravel	Easy driving	Unsteady irregular penetration for successive blows	Dull	None	Rotation is irregular and accompanied by a gritty sound
Boulder and rock	Almost impossible	Little or none	Loud	Sometimes of both hammer and pipe	Dependent on type of formation previously passed through by pipe

COMPLETING THE DRIVEN WELL

There are two ways to terminate the well: above ground, sometimes installed at seasonal, rustic vacation cabins; and the more common, year-round below-ground variety. The choice is yours, but there are several steps to take before you connect your final run of pipe and call the water system finished.

Above-ground termination

To draw water from the well with a hand-pump—either a pitcher pump or one of the special types—terminate the well pipe at any convenient height. If the pipe is too high and cannot be driven farther, you will have to cut and thread it in place. Screw the pump directly onto the end of the well pipe, using an adaptor if necessary. Use plenty of pipe joint compound, and make sure the joints are as tight as possible.

When the pump empties directly over the wellhead, it is best to protect the water supply with a concrete apron around the well pipe (FIG. 12-14). This will keep the surrounding ground from turning to much from spilled water, prevent dirty water from seeping down into the well, and protect users of the well from muddy feet.

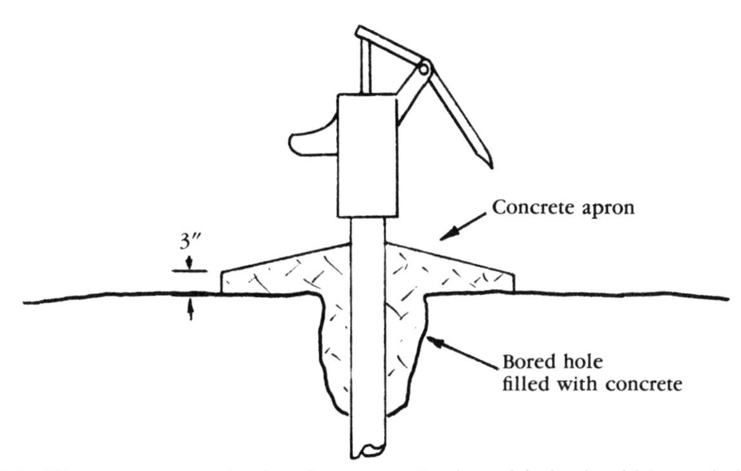

12-14 The space around a hand pump and a bored hole should be sealed with concrete.

Make the apron 3 or 4 feet in diameter and 3 to 6 inches thick, sloped upward toward the well pipe at the center for good drainage. You can use a simple hoop of wood or sheet metal for a form. If the concrete mix is fairly thick, you can easily trowel and shape it into a satisfactory slope.

Underground termination

For a year-round, hidden, automatic water supply, the well termination should be made underground. This can be done by connecting an elbow to the top of the well pipe at some point beneath grade level (FIG. 12-15), or attaching a pitless adaptor (FIG. 12-16). The advantage of the latter arrangement is that you can uncover the top of the well pipe. The adaptor can also be positioned above ground if that is desirable.

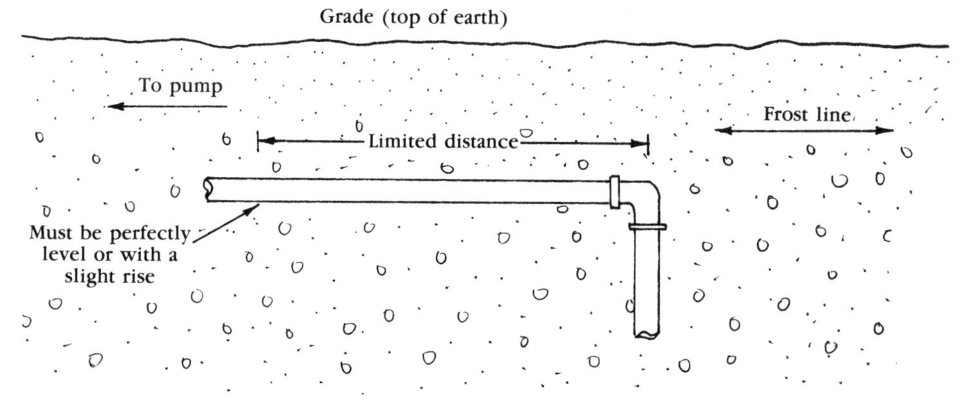

12-15 Terminating a well pipe below the surface of the earth: when a vacuum pump is used, distance to water and distance from well pipe to pump is limited.

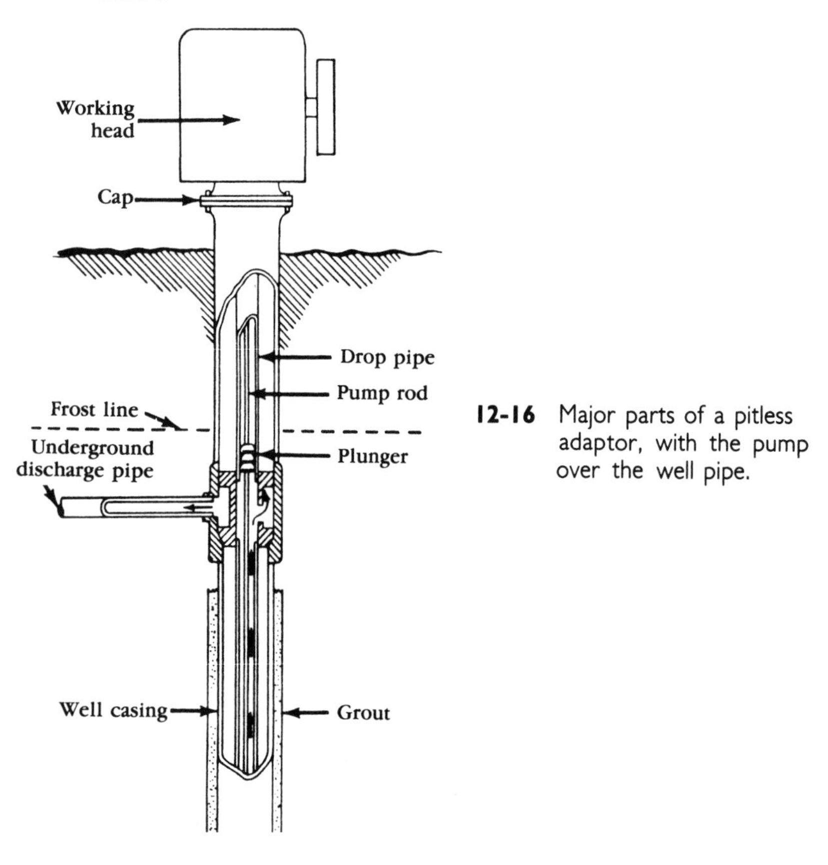

12-16 Major parts of a pitless adaptor, with the pump over the well pipe.

There should be no problem when you install a pumping system that depends upon a pipe or pipes that run down into the well pipe itself, or a system that encompasses a pump motor near or below the water line. When you use a straight vacuum pump of any kind, bear in mind that you will lose vacuum in the horizontal pipeline in addition to the vertical well pipe itself. Therefore, you cannot depend upon a straight vacuum pump to always lift water a distance of 20 feet; the total height will be reduced.

Last pipe or riser

Plan the elevation (height above or depth below the surface of the earth) of the pipe termination before you drive the final length of well pipe or riser. If you are using an underground termination, you need to ascertain the depth of the frost line (if applicable) and dig a hole to below this depth around the pipe, plus another 6 inches for working room (FIG. 12-17).

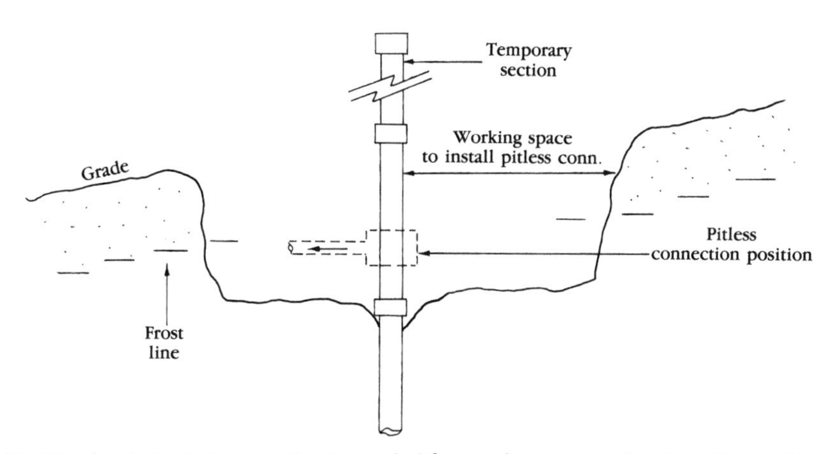

12-17 A substantial excavation is needed for working room when installing a pitless adaptor.

If driving is fairly easy, you can safely assume that you can drive the well pipe another couple of feet without problem. In such cases you can add a temporary additional length of well pipe, drive the well pipe to the desired depth, then remove the temporary piece.

If driving is difficult, you may reach your practical depth limit too soon. When the pipe becomes impossible to drive and yet remains too high for an underground connection, you have two choices. You can pull the well pipe back out of the hole to the first coupling, remove the top pipe length, replace it with another, shorter length, and redrive the pipeline to the desired elevation, or enlarge the hole around the well pipe to give yourself sufficient working room, then cut the pipe and thread it in place. Bear in mind, though, that if you have driven riser pipe, the threads are different than standard pipe threads. Securing the right die and stock to thread the pipe may be a problem. An alternative might be to devise an adaptor that will let you change pipe styles.

TESTING

You can make a preliminary test of your well before you develop and cap it. If the water level is too deep for a suction pump, it saves time and money to know that fact before a pump is purchased and installed. If the flow is grossly inadequate, chances are that development will not increase it sufficiently, which also needs to be considered before you go any further. You might need to take some or all of the steps suggested here for increasing water flow into the well point.

Water level

The water level can be easily ascertained with a float and weight, such as is used for fishing, attached to the end of a line. Lower the float and weight into the well pipe. When the float reaches the water it will carry the weight and there will no longer be a pull on the line. Mark the line with a piece of tape or yarn at the point where it leaves the well pipe. Pull the line up and measure the distance from the tape to the float (FIG. 12-18).

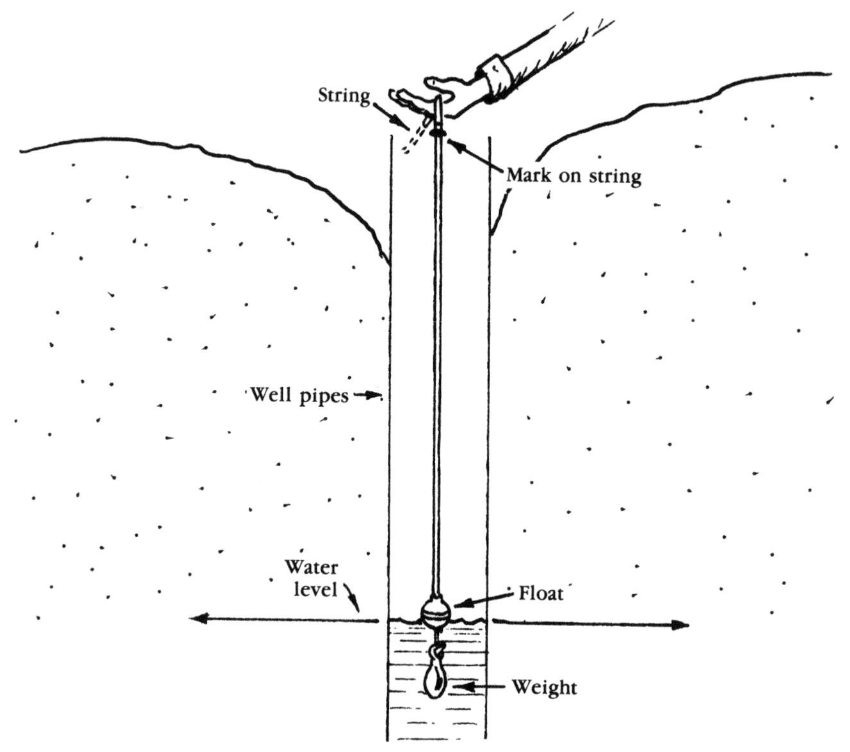

12-18 One way to determine the water level in a well pipe.

Flow rate

Attach a pump to the well pipe, start it up, and let it pump directly into a 55-gallon drum or a series of buckets (have a helper handy to empty the buckets some distance away from the well). Try for a steady, continuous stream of water, even if you are operating a hand pump.

Using a stopwatch, pump for exactly one minute, then calculate the amount of water drawn and make a note of it. Now operate the pump for a continuous five minutes. Quickly pull the pump, drop the weight and float down the well pipe, and measure the water level again. If there has been a drop large enough to measure with any accuracy, make a note of it.

Continue to measure the depth of the water at regular, short intervals, keeping exact track of the time. Ascertain just how much time is required for the water to come back up to its original level. For example, assume that the level before pumping was exactly 20 feet below the top edge of the well pipe. Your one-minute test produced 5 gallons of water. You waited a bit and then pumped a steady 5 minutes and drew 25 gallons of water—the 5 gpm (gallons per minute) rate remained steady. Quickly you remeasured the water level and found that it dropped 5 inches.

Because you know from your records the overall length of the well pipe to the point, you should also know (by subtraction) the height of the water in the well before pumping.

Let's assume that height is 50 inches. If you exclude the quantity of water entering the well pipe through the screen, you can pump water at the previously determined rate of 5 gpm for a total of 250 gallons before you empty the well. This figure is obtained by dividing 50 inches by 5 inches (the distance the water was lowered by drawing 25 gallons at 5 gpm). This equals 10, and 10×25 gallons equals 250 gallons.

Recovery rate

Flow rate alone is not a complete measure of a well's performance. We also need to know the recovery rate, the speed at which water flows back into the well to replace that which has been drawn. Let's assume water entered at a rate of 5 gpm (which is not the case in our example). If this were so, the well theoretically could be pumped without running dry for an indefinite period if the pumping rate were held steady at 5 gpm. But this is not the case. The water level drops 5 inches with every 25 gallons drawn at the 5 gpm rate.

You have to measure the recovery rate to learn just how much you can draw at a steady rate. To do this, give the well a rest to let the water level reach its normal, full height. Then pump to draw the water level down a measured distance. This should be at least a foot, because short distances are difficult to measure accurately with the line and float. Then use a stopwatch to measure the time it takes the water in the well pipe to regain its original level.

Let's assume you pumped the water level down 20 inches (the pumping rate is not important here). This drop represents 100 gallons; each 5 inches of drop, as previously measured, represents 25 gallons. Now time the recharge (recovery) rate. If exactly 40 minutes are required, now divide 100 gallons by 40, which gives you 2½.

You can therefore pump this imaginary well at a rate of 2½ gpm without running it dry. You can also pump in short "bursts" of 250 gallons at any rate if you then wait 90 minutes or so for the well to recover.

Chapter 13

Bored wells

A bored well is made by forcing an auger bit, rotating under pressure, into the earth until a suitable hole has been bored. Well bores ranging from 2 to 32 inches in diameter can be made this way (FIG. 13-1).

Hand-powered augers can only be driven to a depth of about 15 to 20 feet, but powered augers can often go down 125 feet. Boring is best suited to homogeneous formations that contain no boulders or cobbles that must be removed before boring can proceed. The fewest problems are encountered when boring through clay, silt, and sand, or mixtures thereof, that will not collapse behind the auger. From a practical standpoint, bored wells are almost entirely confined to these soil types. Standard boring equipment will not function in hard rock, but there are a very few specialized rigs that will cut through consolidated formations provided they are not too hard.

WHY BORE?

You should consider three factors before you decide between a bored and a driven well, especially if you hope to construct the well by hand. These factors are well diameter, flow rate and packing.

Well diameter

It is usually much easier and faster to bore a 6-inch well than it is to drive a well point of the same diameter. The main reason for driving a 6-inch well rather than, say, a 3-inch one is that you might want or need to install the pump inside the well. Also, when the soil is primarily hard clay, it might be easier to bore than to drive wells of smaller diameter.

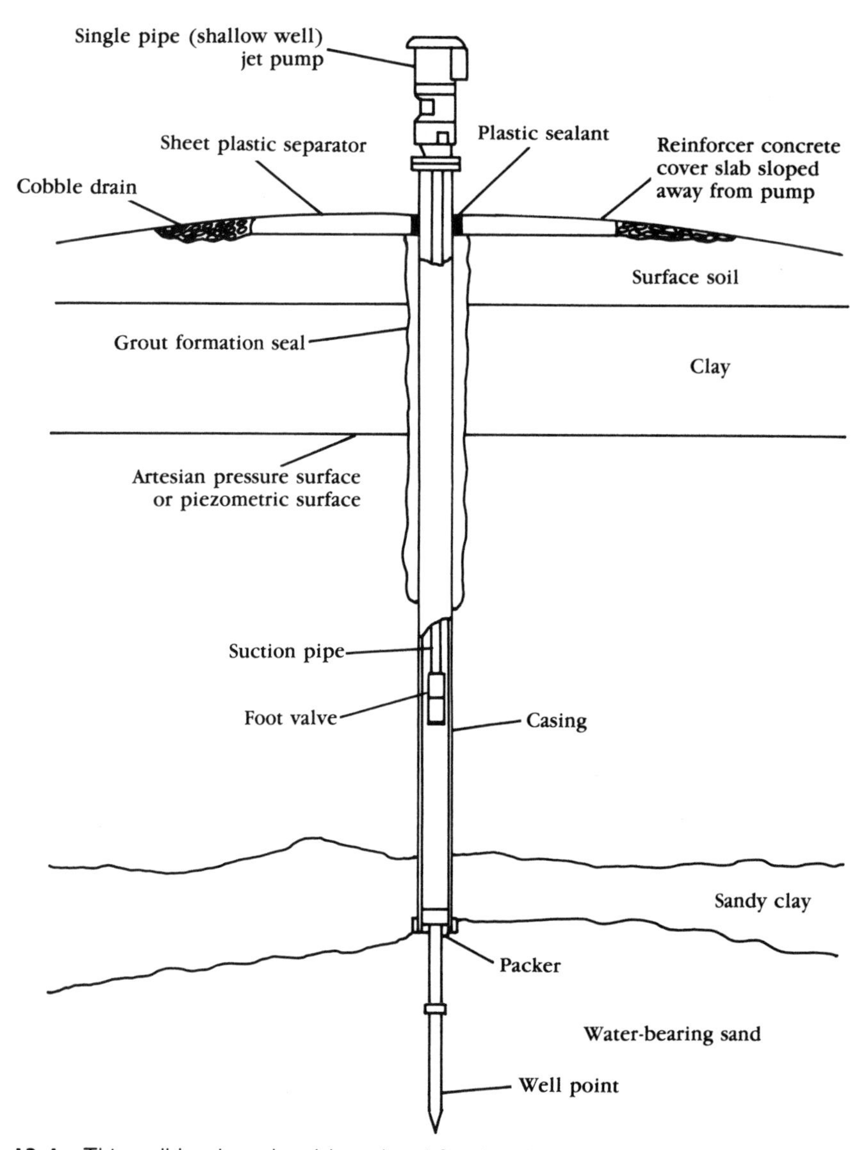

13-1 This well has been hand-bored and fitted with a well point and packer and a shallow-well pump.

Flow rate

It is only natural to assume that by increasing the diameter of the well—meaning the well screen, which is the effective inlet for the well—the flow rate will be increased proportionately. Unfortunately, it doesn't work that way. Doubling the well screen *diameter* increases water input by an average of only 4 percent. Well yield is much more dependent upon the *length* of the well screen.

A rule of thumb is that doubling the length of the well screen will double the yield, all other things remaining equal. Yield actually depends upon six factors: aquifer permeability; drawdown during pumping; diameter of the circle of influence; well screen length; recharge rate; and well screen diameter.

Aquifer permeability relates directly to the relative coarseness of the aquifer material. The coarser the material, the more easily water will flow into the well. Drawdown is simply a measure of the flow rate of water through the aquifer and the quantity of water present in it. Insignificant drawdown means rapid replenishment and an indefinite supply of water at a given pumping rate. The diameter of the circle of influence is an indicator of the rate water enters the aquifer surrounding the well. If, for example, when a well is pumped the water table drops over a radius of several thousand feet from the well—the circle of influence—you know that water is not entering the area as fast as it is being pumped out. On the other hand, if the radius is very small the aquifer is adequately supplied with replenishing water.

If you were to drill and redrill a number of wells of increasing diameter in the very same spot, going from a 4-inch diameter well screen to a 12-inch one, for example, water yield would only increase about 12 percent. The relative cost for a 12-inch well as compared to a 4-inch well would easily be on the order of 5 to 1, hardly cost-effective.

Packing

Packing is the term used to denote crushed stone, sand, or gravel that can be placed between the well screen and the surrounding earth. Under certain conditions packing can effectively increase the effective inlet area of the well screen and so increase well yield. This depends largely upon the composition of the aquifer. It is most often used in fine sand and silt, rarely in coarse gravel or coarse sand, but many experts feel that packing is not worth the effort.

Packing a driven well is very difficult. Packing a bored well is not especially easy, but is done routinely with little problem. Packing techniques are discussed in Chapter 18.

HAND-TOOL BORING

Hand-tool boring is done with a special two-hand auger (FIG. 13-2). If the top layer of earth is soft and loose, the auger will not support itself. Start by digging a hole slightly larger than the auger diameter, down to solid soil, preferably with a posthole shovel.

Using the auger

Force the point of the auger into firm subsoil to start the bore. Make sure the auger is plumb, and rotate the blades until they fill with loose dirt. When you have a full load, withdraw the auger, dump the dirt, and start again.

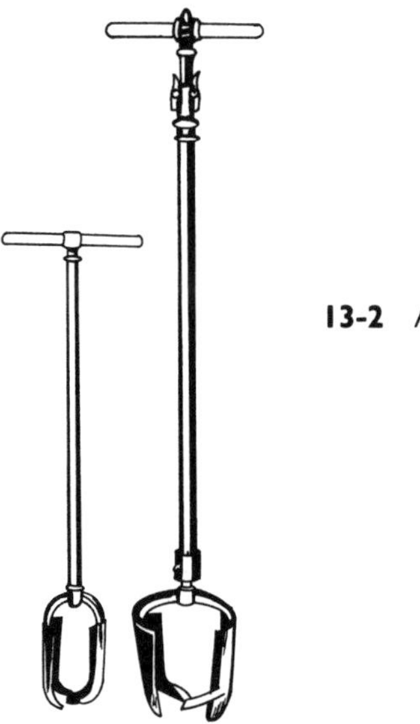

13-2 A pair of hand augers.

Depending on the conditions of the soil, the depth of the auger, and your arm strength, you will eventually reach a point when you can no longer turn the auger with the handle. When this happens, try backing the auger out slightly, turning it at the same time to enlarge the hole and loosen the blades in the earth. When that trick becomes ineffective, extend the handle with a length of pipe to increase your leverage. Be cautious, though, because you can develop so much torque in this manner that you can break the handle or twist the auger shaft.

When you add one or more extensions to the auger, its length will make it awkward or impossible to operate. Provide yourself with a broad, sturdy platform from which to work. When you reach a depth where standing on the platform no longer helps you pull the loaded auger up out of the hole, you need to set up a tripod or derrick with a block and tackle to take the weight of the extended auger.

Cobbles

Small stones will nest into the blades of the auger and be lifted out with the spoil dirt. Larger stones—cobbles or cobblers—have to be dug out one at a time with a spiral or ram's horn auger (FIG. 13-3). Remove the dirt auger and lower the spiral auger gently, twisting it about until you catch the cobble and turn it up into the spiral. Then you can lift it out of the bore.

Should you encounter a large rock, there is no point in trying to continue. All you can do is start another bore elsewhere and hope for the best. There is no way to remove a rock larger than the bore hole using hand tools.

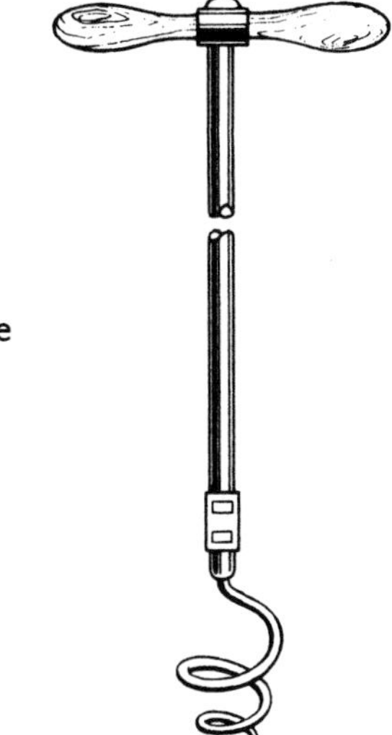

13-3 A spiral auger used to remove stones from the bore hole.

Working in collapsing soil

Making a bore hole in soil that will not close in upon the opening as you remove the auger is merely a matter of straightforward work. But when the soil collapses into the bore behind the auger as soon as you remove it to clear the hole, your problems are compounded. You can abandon this technique and sink the well by jetting it, which I discuss in Chapter 14, or you can case this bore as you proceed.

In firm soil, you case the bore hole after you have drilled it to the desired depth. But if your soil is collapsing, you must case the well as you bore it (FIG. 13-4). As you can imagine, casing while boring is much more difficult.

Select a steel casing with an internal diameter at least ½ inch larger than the external diameter of your auger. You need this clearance to

13-4 Pipe tongs are a large chain-type pipe wrench.

prevent the auger from "freezing" inside the casing. You can cut jagged "teeth" into the bottom end of the casing to make it drive a bit easier if you wish.

Bore a hole a foot or so into the earth. Remove the auger and position the casing vertically in the hole. Place a temporary protective cap, such as a pipe cap or a chunk of 2 by 4, over the top of the casing. Drive the casing down into the earth with a sledge as far as it will readily go, keeping it perfectly plumb meanwhile. Sometimes a small amount of water poured into the hole and also around the outside of the casing will make it easier to drive, but too much can turn the soil into mud that the auger can't lift out.

When the casing has reached its practical drive limit, check to make sure it is plumb; if it is not, straighten it. Then slip the auger down inside the casing and remove some more soil. Remove the auger and spoil dirt and drive the casing down a bit further. Repeat the process until you've reached the desired well depth. You can turn the casing with a pipe or chain wrench (FIG. 13-5) as you drive it. Two workers can bore and turn the pipe simultaneously, which often makes the job go faster and more easily.

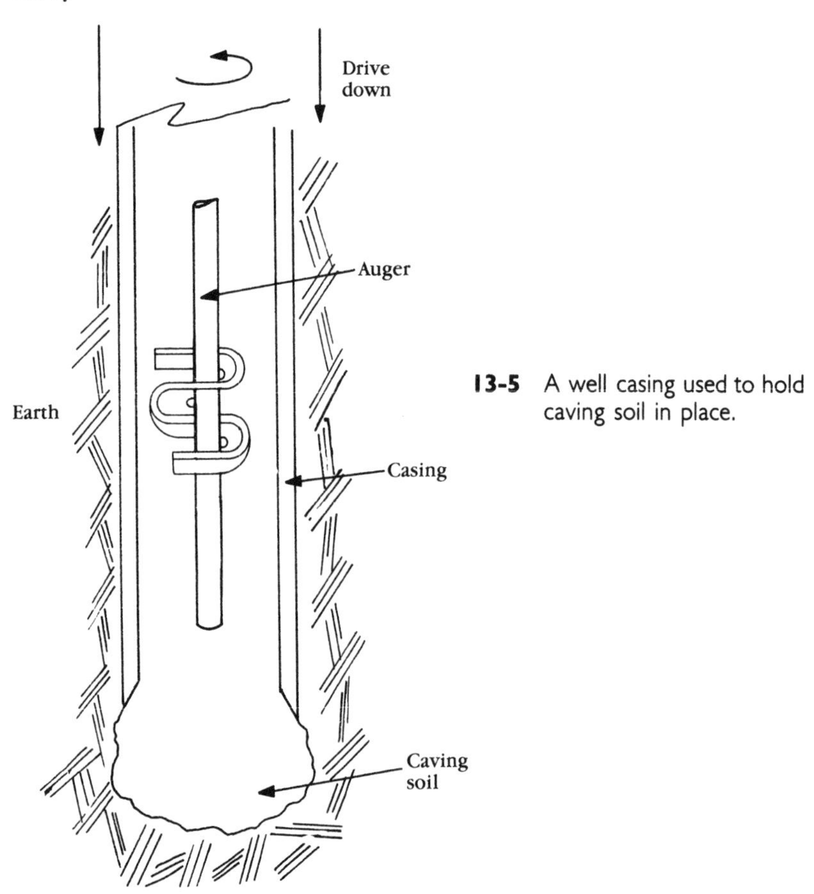

13-5 A well casing used to hold caving soil in place.

When the soil in which you're boring does not collapse, unless it is firm clay or a mixture of clay and sand or silt right to ground level, dig a pilot hole by hand first to get rid of the soft topsoil. Then boring is just a matter of turning the auger and lifting out the spoil.

Your well is completed by setting a well screen at the lower end of the casing. This is covered in Chapter 17.

Casing the bore

When casing in collapsing soil, use an auger with a smaller diameter than the inside diameter of the casing, so you can bore down inside the casing. When you work in noncollapsible soil select an auger that is larger by at least 1 inch than the outside diameter of the casing pipe.

Bore the hole in the manner described earlier, and remember that you can't case a crooked bore. When the bore has reached its final depth, insert the casing sections. You can join them either by electric arc welding or by connecting them with threaded couplings. Precut the last length of pipe as required to terminate the well at the desired height.

If you plan to pack the well the bore should be twice the planned packing thickness plus the outside diameter (o.d.) of the casing pipe. For example, if you want to surround a casing having an o.d. of 6 inches with a 2-inch layer of packing, you should bore a hole with a 10-inch diameter (FIG. 13-6). The o.d. of a pipe depends upon its type, trade size, and wall thickness and will vary from one kind and grade of pipe to another. To be certain of the o.d. of the pipe you plan to use, measure it.

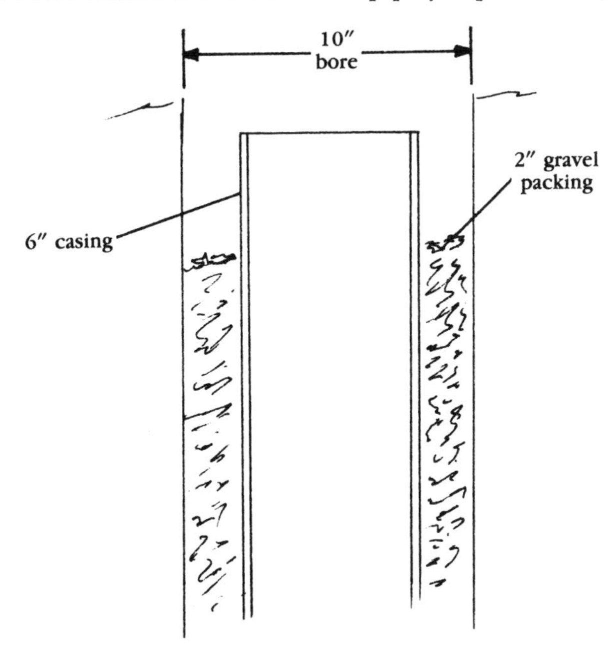

13-6 For a packed well, make the bore twice the packing thickness plus the outside diameter of the casing pipe.

POWER AUGERS

The two kinds of power augers are *continuous* and *bucket.* Like a wood boring bit, the continuous auger is made in a spiral form that digs itself into the earth as it turns, bringing the spoil dirt upward and depositing it beside the hole. The bucket auger consists of a bucket with an open bottom, to which teeth are fastened. The bucket is attached to a rigid stem, square or hexagonal in cross section, called a *kelly.* The kelly passes through a similarly shaped hole in the center of a power-driven rotating table. When the table turns, so does the kelly, yet the kelly and bucket are free to be raised and lowered.

The continuous auger typically can bore to a depth of 18 to 20 feet with an auger 4 inches in diameter, given good conditions. The bucket auger can do much better. Working in an unconsolidated formation, it can bore at a rate of 30 to 40 feet per hour, sometimes as much as 60 feet. When conditions permit, commercial drillers can bore five or six wells a day. Bore hole depth is limited by the length of the kelly and the number of extensions that can be added, but can be as much as 125 feet. Maximum diameter is typically 30 inches, which enables the bore hole to be lined with 24-inch i.d. (inside diameter) concrete pipe.

The bucket auger works best in unconsolidated formations, because they will not collapse behind the bucket. When slight or mild caving is encountered, many rotary bucket-drill operators fill the bore hole with water and depend upon the hydrostatic pressure to keep the walls from collapsing.

This might seem odd in light of my assertion that too much water in a hand-augered bore will turn the soil to mud that the auger cannot lift clear. The difference is partly due to the rotational speed of the bucket auger and partly due to the shape of the bucket. The bucket acts as a scoop. When it has scooped up the mud and continues to turn rapidly, the mud remains in the bucket and can be withdrawn. This technique works best in sand.

Casing a rotary-bucket bore hole

Casing is necessary when the soil is so soft that it will not stop collapsing even when the bore hole is filled with water. The standard bucket is then replaced with one that has "expanding" teeth on the bottom edge. The purpose is to bore a hole several inches larger in diameter than the bucket itself.

The hole is bored as deep as can be done without the walls caving in. Then a length of casing is inserted into the hole and positioned vertically. The special bucket is lowered into the casing to the bottom of the hole and rotated until enough soil has been removed to lower the teeth below the bottom of the casing. The teeth expand outward and cut a bore larger than the casing diameter. The casing moves downward of its own weight, or can be easily pushed or driven down.

Casing a bore during drilling increases the cost of well material and labor. If you prefer, the casing can be left in place and used as the well casing or pipe. Or you can install a second pipeline inside the casing to be used as the well pipe.

Boring procedure

The site selected must be easily reached by the truck-mounted drilling rig, and sufficiently level, or capable of being leveled, that the machinery can also be leveled. The rig is positioned on-site with the back end positioned over the bore location. Hydraulic outriggers are lowered to lift the machine off the ground and level and stabilize it. The mast or derrick is hydraulically raised and locked in place, perfectly plumb.

The telescoping kelly is then lifted into position and one end is lowered through the hole in the rotatable table. This is part of a large *ring gear* driven by a *pinion gear*, which in turn is attached to the *drawworks*. The latter is the term used to describe the engine and attachments that drive the table and also lift ("draw") and lower the kelly and bucket.

The bucket is then attached to the kelly and lowered until it rests upon the ground. The table is rotated and the bucket digs into the earth. When it is full, the kelly is retracted, lifted free of the table and brought to one side of the rig where the soil is dumped for later removal.

Soil sampling

One big advantage that bucket-boring, or even hand-boring, a well has over other methods is that every time you bring up a bucket of earth you bring up a sample of the formation you are boring into. Thus you know almost immediately when you have struck the water-bearing formation and also when you have unwittingly gone past it. The latter situation poses no serious problem because the lower portion of the bore hole can be sealed with gravel, crushed rock, or even concrete to whatever depth is required.

Casing an open-bore hole

After the hole is bored, the casing is lowered into place with the aid of a *trip line*, a cable with an automatic release mechanism. This is done one section of casing at a time. The first section is suitably pierced to permit the entrance of water. When the well is to consist of but a single casing used as a well pipe, the bore hole is made oversized by several inches. The space between the casing and the walls of the bore hole lying within the aquifer is packed with coarse sand, gravel, or crushed rock.

Chapter 14

Jetted wells

Jetting is a method of constructing a well, like the one shown in FIG. 14-1, using a strong flow of water to bore a hole in the earth. If you were to connect a garden hose to a source of water under pressure, place the hose nozzle against the earth, and turn the water on, the jet of water would bore a hole. If you were to continue pushing the hose nozzle into the hole, the force of the water would continue to loosen the soil at the bottom of the hole and the movement of the water would carry the spoil dirt up and out of the hole in a slurry. As soil came up, you could lower the hose nozzle deeper and deeper.

This is the basic principle behind jetting a well. In actual practice the garden hose is replaced with a rigid pipe. The flow of water can be aided by a flow of compressed air. The pipe can be rotated and it also can be driven. The jet can be used to form a hole for itself or for a well casing.

Jets of water and air are also used with rotary and percussion drilling. Here we will be looking at well-drilling techniques that depend mainly on the flow of water and sometimes air for excavation.

MAJOR PARAMETERS

No specific rules outline the quantity of water required for jetting a well or the rate of penetration possible. The larger you want the bore hole to be, the more compact the soil is, or the deeper and faster you want to go, the greater the volume of water and amount of pressure needed.

Water at 50 psi (pounds per square inch) is generally adequate. However, this is no reason not to try jetting with whatever water pressure is available, whatever the source. The only result a lack of adequate pressure at the lower end of the jet has is a failure to lift the spoil dirt out of the hole. Low pressure will also limit the depth to which your jet will effectively work and limit the penetration rate.

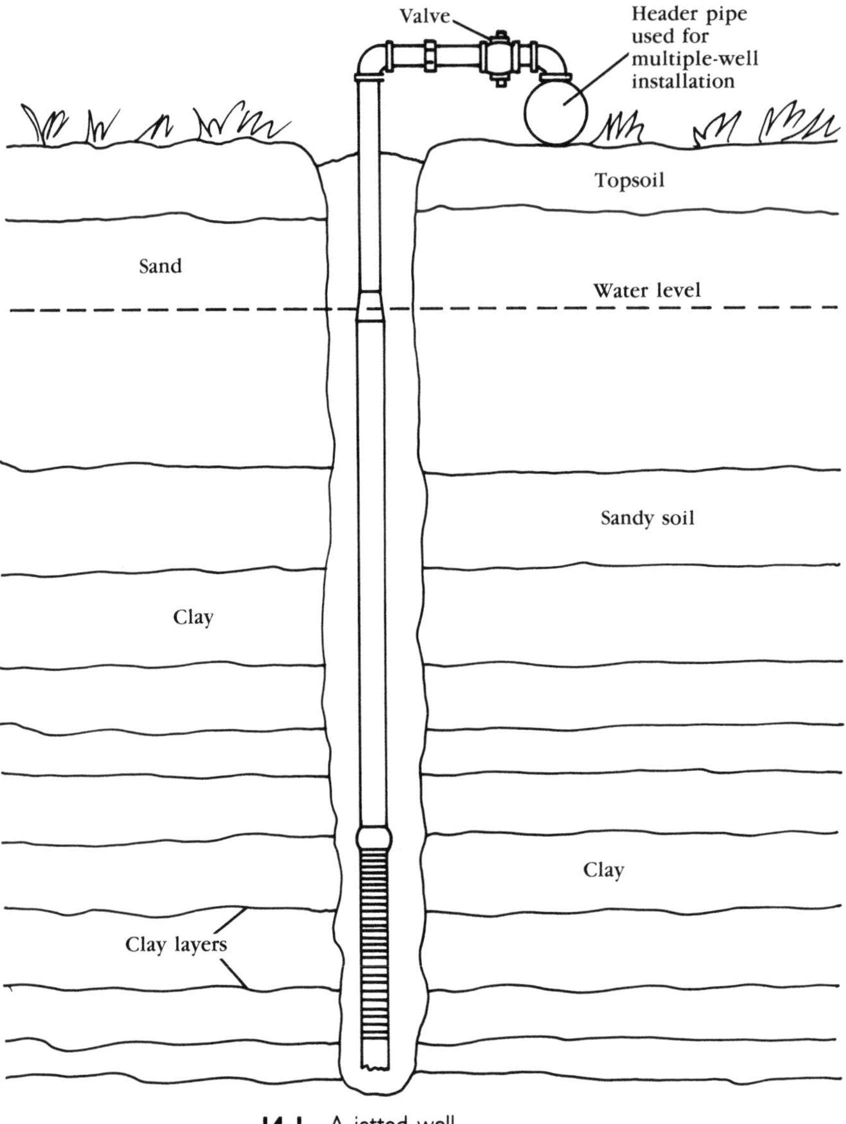

14-1 A jetted well.

If you are going to rent a pump to provide the required water pressure, you must first determine the quality of the water you will use. If it is fresh and clean, you can use any kind of pump with sufficient capacity. If you will use recirculated water, the usual arrangement when there is not enough clean water at hand, you must use a centrifugal pump. An ordinary piston pump would soon be ruined by the abrasive sand and silt in the pumped water.

A volume of some 50 to 100 gpm is generally sufficient for most work. The most practical approach is to make a test run with what you have.

Pump and driver

For best results, the electric motor or gasoline engine driving the pump should be matched to it. If the pump is too large, the driver will be overloaded. If too small, it will be rotated beyond its designed range, cutting its efficiency. In most cases the units will be properly matched when you rent or purchase a pump and drive set. Unless the water supply source is higher than the center line of the pump, the pump will need to be primed before it will begin pumping.

Bore hole diameter

The diameter of the hole the jet stream makes is always larger than the o.d. of the jetting pipe. Typically, a 1-inch pipe will produce a 2-inch hole. A 1½-inch pipe will create a 3-inch hole, and a 2-inch pipe will make a hole 4 to 6 inches in diameter.

Jetting speed and depth

Some drillers working with simple jetting equipment, but with plenty of water under high pressure, report that they can jet a small-bore hole to a depth of a dozen feet in a few minutes. In heavy formations, jetting alone will not produce penetration, or will do so only at a very slow rate. Under such conditions, penetration is only achieved at a reasonable rate when the jet pipe is also driven.

The jetting system is most often employed for shallow wells, 25 feet or so in depth. Soil structure permitting, jetting is an even faster way of sinking a well than driving it. Maximum depth with jetting alone is limited, and depends upon the nature of the formation. With moderate driving, on the other hand well casings have been sunk as deep as several hundred feet.

The deeper you go, the greater water pressure you need to scour out the hole and force the soil slurry to the surface. The weight of the water alone is 0.43 pounds per square inch per vertical foot; to lift water 1 foot you need a pressure of 0.43 psi. Thus, 44 psi is required to bring just the water to the surface of a 100-foot bore, not taking into account friction loss (of which there is plenty). More force is needed to bring up the spoil dirt. Sand usually requires less pressure but more water, clay just the opposite. Sometimes as much as 150 psi is needed just to move clay and lift gravel a modest distance.

If the single centrifugal pump you have will not do the job, you can connect two of them in series. For satisfactory results they must have similar capacity; otherwise, arrange the larger unit to feet the smaller.

MAKING YOUR OWN JETTING EQUIPMENT

You can readily construct a jetting tool, sometimes called a *lance* or *jet lance*, from a length of ordinary water pipe with an interior diameter of 1 inch or more. Flatten one end to make a slot-like opening about ¼ inch wide. Grind the exterior edges of this end down to a blunt chisel point (FIG. 14-2). The other end of the pipe should be threaded. Screw on a 90-degree elbow, and turn a 12-inch nipple into that. Fasten the water supply hose to the open end of the nipple, using whatever adaptor fittings are necessary. Figure 14-3 shows the arrangement.

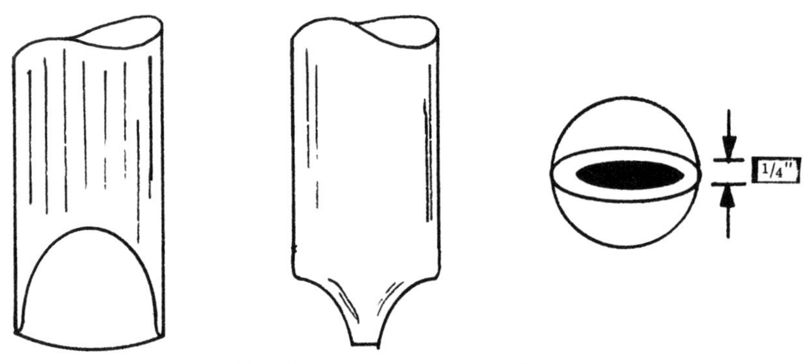

14-2 The tip of a homemade jet lance.

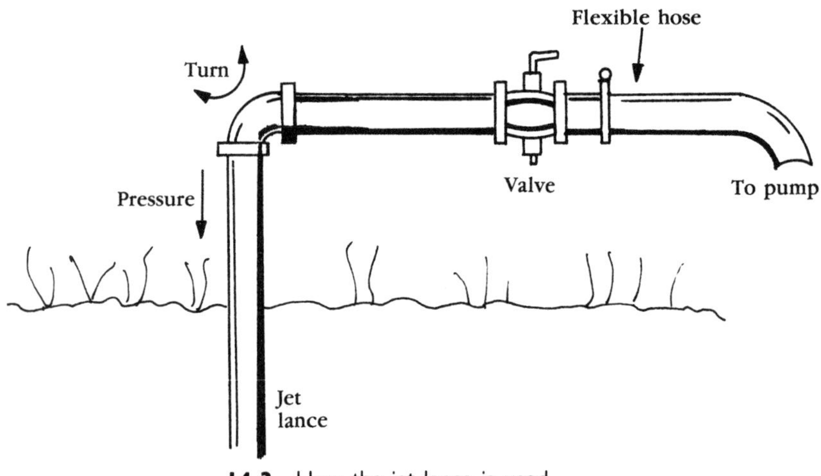

14-3 How the jet lance is used.

To penetrate deeper into the earth than a single length of pipe will permit, unscrew the elbow, fit another length of pipe, and replace the elbow. You can repeat this as often as necessary to the depth as you want, or as deep as water pressure, available water and formation conditions will permit.

An alternative to the homemade jet lance consists of the same water pipe, to which is threaded a commercially manufactured bit (tip). When selecting a bit, make sure the diameter and thread pattern match that of the pipe you have or plan to use. If the bit has an odd or special thread, you will have to secure similar pipe and other fittings, or adaptor fittings, which could be troublesome, expensive, or both.

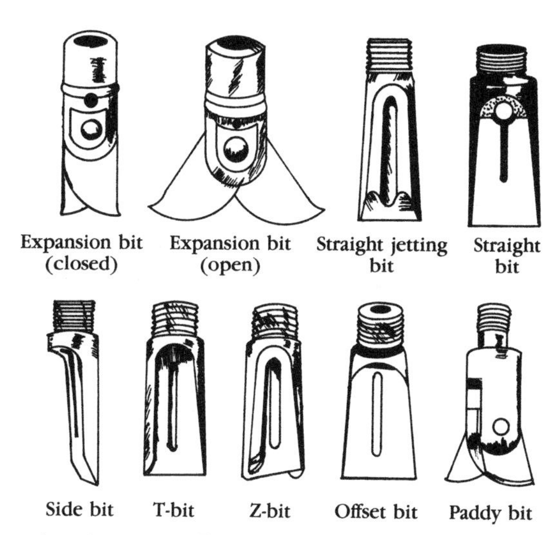

14-4 Commercial jet-lance bits. The expansion and paddy bits are designed to be slipped through a casing or well pipe, then opened up.

Commercial bits include straight, straight-jetting, side, T-, Z-, and offset bits (FIG. 14-4). Expansion and paddy bits are designed to be slipped through a pipe casing, then expand and jet a hole considerably larger than the casing, so the casing can more easily be jetted in place.

Another tool that can be made in the home shop is shown in FIG. 14-5. This device can be used to jet and install the well casing in one operation, using both air and water.

USING THE JET LANCE

The initial problem with what might be called "hand equipment" is making certain the lance is perfectly vertical during the first few feet of penetration. This is not difficult if there are two workers; it often helps a single worker to start by digging a hole a foot or so deep and constructing the simple positioning guide shown in FIG. 14-6.

Position the lance vertically with the tip resting upon the ground. Turn the water on part way and rotate the lance back and forth through a short arc to direct the water stream in an even flow. When the lance tip has penetrated 6 inches or so, complement the rotating action by gently raising and lowering it a little, again to produce a more even flow.

As the lance continues to bore into the earth, continue these motions. Eventually the water backflow will have little color and debris. This indicates that either the pressure is too low or the formation is too hard. Try increasing the pressure. If that does not materially increase the penetration rate, try driving the jet lance. Raise the lance about a foot and drop or thrust it back down. Turn the pipe slightly and repeat. Driving with a guided weight (FIG. 14-7) might also help.

When you reach a point where vigorous driving and full water pressure no longer advances the lance into the formation, you have

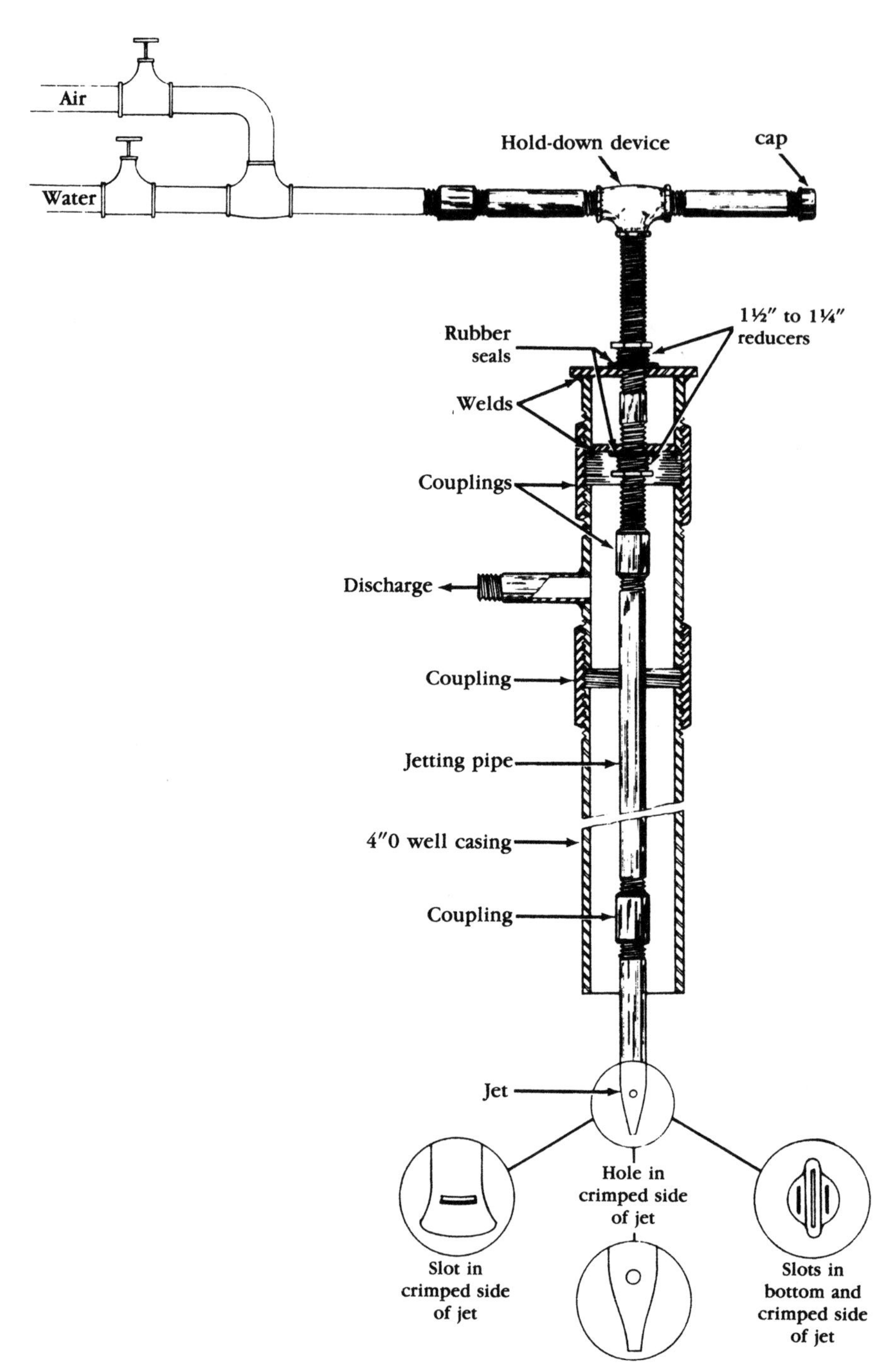

14-5 A jetting arrangement that uses compressed air as well as water for faster drilling.

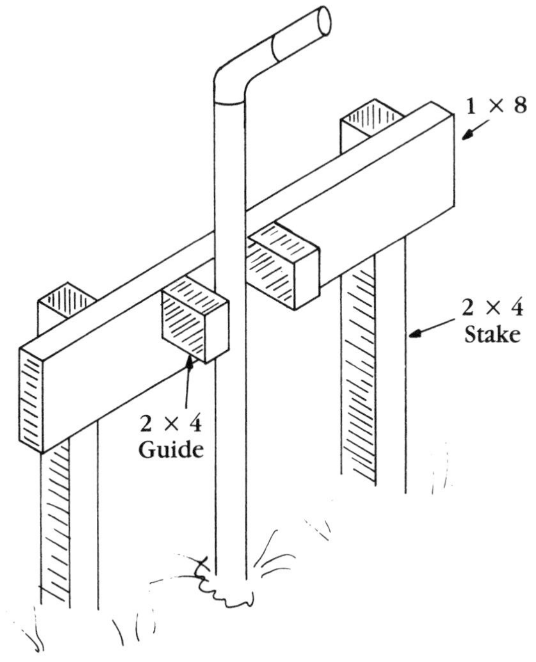

14-6 A simple guide can be built to hold the jet lance in a vertical position during the jetting process.

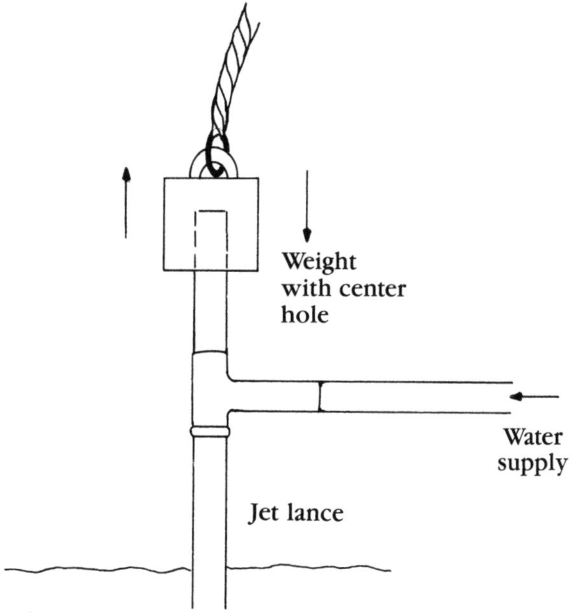

14-7 A weight can be used to drive a jet lance while the flow of water washes soil from beneath the tip.

either reached bedrock or a boulder, or you are at the limit of your equipment. If the water backflow slows appreciably or stops but the pump continues to operate normally (even running a bit harder than usual), you will find that the orifice in the lance tip is plugged.

Stop the pump and remove the lance, but be careful to keep the bore hole full of water. Clean out the orifice. If clogging becomes a nuisance, drill a number of ⅛-inch holes in the lower end of the lance. Position them at the outside of the pipe so the water jets travel horizontally at right angles to the pipe axis.

Keep close tabs on the debris brought to the surface by the stream of water. The nature of the debris will help you determine when you have reached water-bearing sands. When you have penetrated to a sufficient depth either to center your well screen in the thickness of the aquifer (the water-bearing sand) or place it several feet below the aquifer's top edge, you are ready to complete the well.

COMPLETING THE WELL

Once you have jetted the preferred distance into the formation, turn down, but don't turn off, the water flow. Keep a slow, gentle stream entering the bore hole. Then prepare the well screen and the well pipe. The same type of point described in Chapter 12 is attached to a length of well pipe and the joint liberally doped and made up tight. Place the other required lengths of pipe in a nearby handy spot, with the joints doped and a coupling threaded tight on each length.

Remove the jet lance from the well and slip in the well point. One by one as you lower the pipeline into the well, add the subsequent pipe sections and make them up tight. If you have a long length of pipe to make up, arrange some sort of clamp to take the weight of the line as you attach more pipe, and to make the lowering process easier.

A 10-foot length of 1-inch pipe weighs about 17 pounds. A length of 2-inch pipe weighs about 37 pounds. If you are down to a modest depth of only 40 feet and running a 1-inch line, you will have 68 pounds or so after you couple the last length. With 2-inch pipe, you would be trying to hang onto and maneuver 136 pounds. Good luck!

Working in collapsing soil

The first indication that the soil is caving in on a simple jet lance is that the sides of the bore hole collapse some even while the water jet is flowing and the lance is in the bore hole. Another indication might be the closing of the bore hole under the end of the lance. In that case, when the lance is rotated and raised and lowered, perhaps even driven, there will be considerable penetration at first. But on the next sequence there will be suddenly little or no penetration. This could indicate that the lance has reached bedrock, but it could also be due to the walls of the bore hole caving in and filling the bottom of the hole.

Keep the bore hole full of water at all times. During any time that the lance is removed from the hole, direct the hose to keep the hole full.

In many instances the water level in the bore hole will drop rapidly once the water is shut off.

Another solution to the caving problem consists of holding a well point and well pipe alongside the jet lance (FIG. 14-8). Both pipelines are moved up and down together so the jet stream opens a path for the well point. Even though the bore hole walls will collapse, they do so upon the well point. After both pipes have been jetted to the optimum depth, the lance is removed. Removal will be made easier by continuing to let the water flow as you raise the lance. To keep unwanted soil from coming up out of the bore hole with the lance, reduce the water pressure.

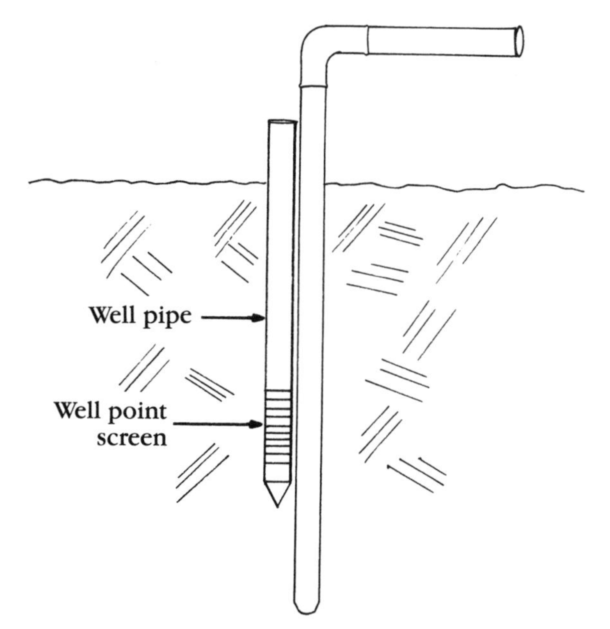

14-8 A jet lance can be used to wash a well point into place.

Self-jetting well points

Self-jetting well points consist of a screen with provisions for attachment to a riser pipe at one end and a reinforced point at the other. But unlike a driven well point, the bottom end has one or more openings leading to a spring-loaded one-way valve that permits water to flow down and out through the holes but will not allow water to enter.

The self-jetting well point has several advantages, but there is one serious drawback that makes it far from being first choice when selecting fast and inexpensive well construction equipment. It requires a second pipe that is slipped down the length of the outer pipe (FIG. 14-9). If just a single pipe was used, water pumped down to a self-jetting well point would flow out through the screen. Instead of high-speed water jets

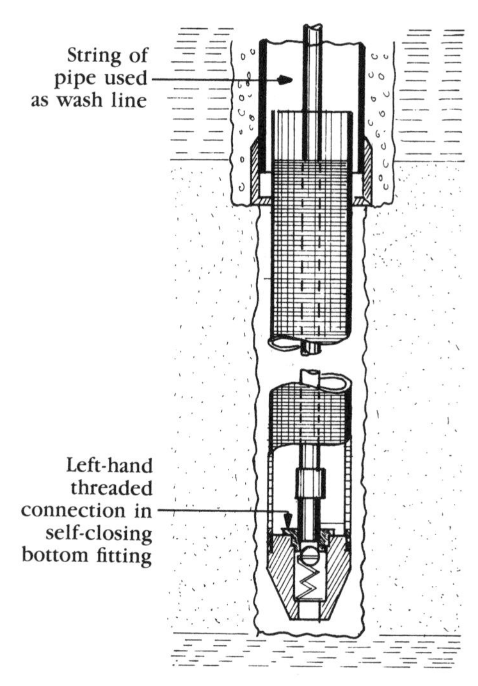

14-9 Inside view of one type of self-jetting well point. After the jet has washed its way down to the preferred depth, the inner pipe is unscrewed.

capable of cutting into the earth, there would be only a large volume of water slushing through the screen.

A self-jetting well point attached to its pipe, with a second inner pipe attached to the water supply, is operated just as is the single-pipe jet lance. Both pipes can be lowered and raised as a unit, or the inner pipe alone can be driven in the same way. The outer pipe, along with the point and screen, can be rotated gently to and fro to keep stones from catching in the mesh and ripping it open, and to make the jetting process easier.

Driving a jet lance

To speed the rate of penetration and possibly to pierce dense clay formations, you can apply more downward force to the lance than is possible by using only the weight of the pipe. When you have a long string of pipe in the ground, driving the lance with a falling weight is much easier than lifting the entire string and dropping it.

You can install a tee fitting on end on top of the lance in place of the elbow. Use the center of the tee for the water supply, and cap the top end so you can drive the pipeline with a sledge. Also, any of the methods described for driven wells will work.

WASHING-IN AND DRIVING A CASING

If the formation will not remain open while you jet it, or caves in when you attempt to insert the well pipe, you have little choice but to install a well casing. A casing is also used when a well pipe diameter is required that cannot be easily jetted into place. And when jetting alone cannot penetrate with reasonable effort, a casing can be installed; it can be driven with much more force than can a jet lance.

Select a casing diameter with the preceding restrictions in mind and also consider the outside diameter of the well screen to be used. The well screen should be of the telescoping type; it is telescoped (slipped inside) the casing, slid to the bottom, and positioned just beyond the lower end of the casing. The upper edge of the screen is then expanded to lock in place (discussed more fully in Chapter 17).

To wash in a casing dig a small hole in the earth to facilitate starting, or bore down a few feet with a hand auger. The casing is set vertically in the starting hole. The jet is set inside the casing, the water is turned on, and the casing is washed-in by eroding the soil around it and gradually lowering the pipe. When the casing has reached the desired depth, the well can be completed. To facilitate casing penetration, teeth are often cut into the leading edge of the casing. As water pressure is applied the casing can be rotated with a chain wrench, making the teeth act like a crude hole saw.

A drive shoe, which is a steel ring with a sharp edge, is fastened to the lower or lead end of the casing. A striker cap is placed over the top end of the casing to take the blows of a sledgehammer or falling weight. When water is used in concert with a driver, a setup like the one shown in FIG. 14-10 can be used.

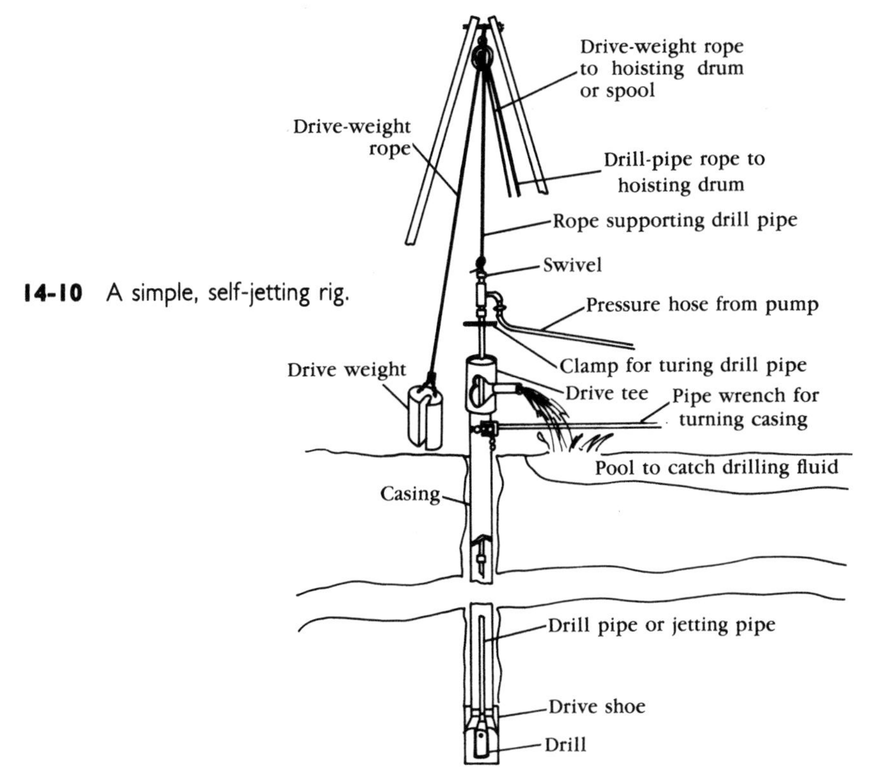

14-10 A simple, self-jetting rig.

Other arrangements are also possible. Bear in mind that the force must strike only the casing. The pipe carrying water to the jet, sometimes called the *drill pipe*, must be free to be raised, lowered, and turned as required.

Chapter 15

Dug wells

A dug well is simply a hole in the earth that is lined with brick, stone, tile, concrete block or steel. The lower portion of the lining, which lies within the aquifer, is pierced to allow water to enter. The upper portion of the lining is watertight.

Only water from the aquifer can be permitted to enter the well; surface water and seepage through the zone of aeration is blocked. When a bucket was the means of drawing water, the walls of the well were built up 3 feet or so above ground level. A small roof was often built over the well, the supports were used to anchor a windless and rope arrangement, and sometimes a removable cover was also placed over the well top.

After the pitcher pump was developed, well walls were commonly terminated just above ground level and the well was covered with planks, slate, or concrete. Modern dug wells can be sealed at or below ground level when used with a modern pumping system, completely hidden from sight.

EVALUATION

If you want a rustic, romantic touch for your yard, a dug well with high stone walls, roof, and windlass might be your answer. If you are looking for an economical source of potable water, a dug well is rarely the answer, especially if you need a substantial, dependable quantity. Laying up brick, block or stone in a hole in the ground is not very difficult, though it is physically demanding. The circular design of the well provides lateral stability that is backed by the surrounding earth. Neatness of construction is of little consequence, although adequate workmanship is required. It would seem, then, that all you need to do is grab your pick and shovel and set to work. This is the way they dug wells years ago and still do in many countries. And this is the way many well diggers have died.

You cannot safely enter a hole in the ground that is deeper than your waist and of relatively small diameter unless the sides of the excavation are shored up, curbed, or braced (the terms are interchangeable, the practice is essential). There is no way to tell by appearance whether the soil might cave in or not. Even with modern construction, safety rules and good equipment, workers are killed regularly by excavation cave-ins. It's not a nice way to go.

A dug well, despite its seeming simplicity, is not just a lined hole in the ground, it is a small-scale engineering project. Unless you take the proper precautions, your dug well could be a large-scale disaster.

There is only one advantage in having a dug well (usually not enough to offset its disadvantages), and that is its storage capacity. A small-diameter pipe well can hold only a few gallons of water, but a dug well is a reservoir. If the internal diameter is 4 feet and the well contains 10 feet of water at a stable level, 940 gallons of water is immediately available. Having a similar steel or concrete storage or cistern tank on the premises would be problematical and expensive, but could be arranged and the tank can even be connected to a bored or driven well. The value of a dug well lies in having a reserve of water against temporary drought and for fire fighting purposes.

Labor

Most of the work in constructing a dug well lies in the digging itself. Difficult to visualize is the huge quantity of spoil dirt that has to be removed and hauled off. A typical dug well might require an excavation 7 feet in diameter and perhaps 20 feet deep, which represents an enormous amount of dirt and rock that has to be hauled of the hole a bucketful at a time. The block, brick, or stone you use to wall the well must be lowered into the excavation. The entire process is slow and laborious.

Material

You can wall a dug well with brick, standard or curved concrete block, rubblestone, precast concrete sewer pipe sections, or large-diameter stackable rings. Often the quantity of materials needed cannot be accurately determined until after the well has been dug, though you can get a estimate easily enough. You will need a quantity of mortar for laying up brick or block. If you pack the well, size-graded gravel will be needed. You'll also need a number of 2 by 8 planks about a foot longer than the planned excavation depth, as well as steel reinforcing hoops or more planking to make braces (FIG. 15-1). If the excavation is deeper than the length of the planks available, a second set will be needed. The preferred wood is spruce, with oak next best. Fir splinters too easily, hemlock is okay if clear of knots, poplar and the pines are too soft.

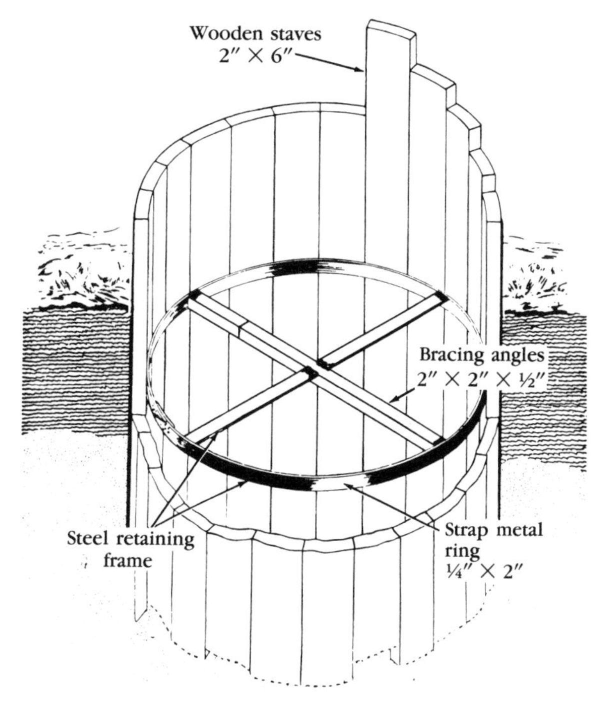

Note: Retaining frame spaced 5 feet apart for average ground formations.

15-1 The retaining ring is made of steel. The staves or planks are driven into the earth one at a time.

Equipment

You will need a pick and shovel, ladder, and perhaps a posthole bar or crowbar for digging. A tripod arrangement, rope fall or electric hoist, and one or two buckets can be used to raise spoil dirt and lower materials. To hand-drive the planks you will need a maul; power-driving requires special jack-hammering or pile-driving equipment, an air compressor and a tripod or derrick. Once you reach water, you will need a high-capacity pump to keep the excavation dry while you continue to work.

DECISIONS

When you site your well, you must consider: presence of water; relation to point of use (barn, garden, house); distance from possible or probable contamination source; appearance, if the well or well house will be above grade; and convenience of spoil dirt removal and disposal.

A test well can indicate whether digging a well at a particular spot would be worthwhile. Drive a small-diameter well down to the top of the aquifer, then deeper to determine whether the aquifer is thick enough for the proposed well.

Remember that the initial well yield can usually be multiplied (some report a nine-fold increase) with proper development. And well yield is directly related to well screen height. In a dug well, yield is directly proportional to the vertical pierced-area height of the lower section of the well wall. Pumping should not bring the static water level below the pierced section, so plan to place it deep enough into the aquifer.

Diameter

Well diameter should be based in part on the quantity of water you want to store in the well. This refers to the water volume when the well is at rest, and does not include the volume taken up by the well lining materials. Figure 15-2 shows the relationship between diameter and yield.

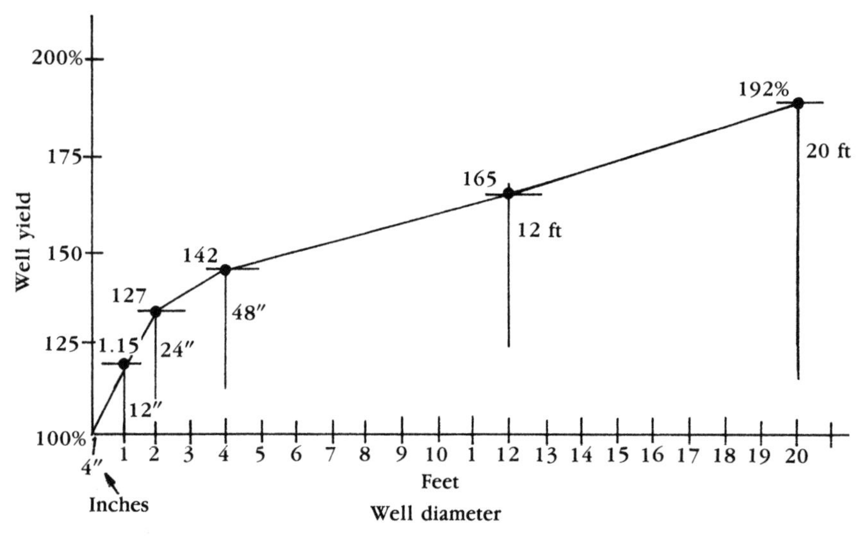

15-2 The approximate relation between well diameter and yield.

If you plan to line with brick or stone, there is no special advantage to making the well diameter any particular size other than large enough to allow sufficient working space inside and around it (FIG. 15-3). If you use concrete block, short blocks allow laying to a smaller radius more easily than standard blocks. Sewer block, which is curved, establishes the inside diameter of the well by virtue of its own arc. If the lining will be precast concrete pipe or rings, the outside diameter of the material plus a foot or so for working room will set the well diameter.

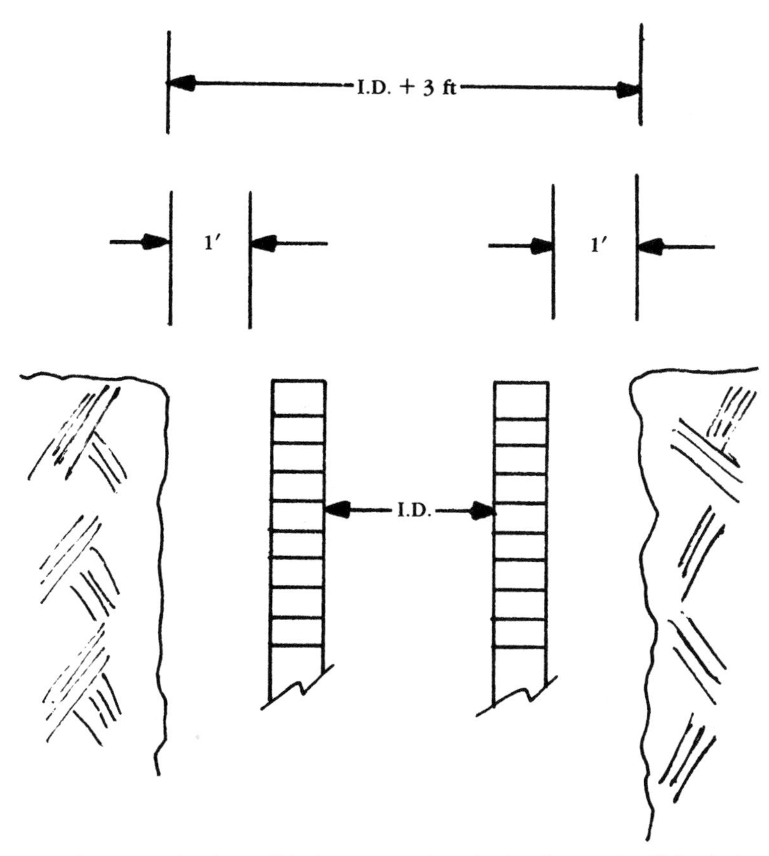

15-3 When you dig the well hole, you must make its diameter sufficiently large to provide at least 1 foot of clearance between the well wall masonry and the surrounding earth, as working space.

Well Termination

You can terminate the well lining about 3 feet above ground and leave the well open but partially protected by a small shingled roof. If you use concrete pipe, the above-grade portion of the wall can be veneered with stone, brick, or tile. Or you can terminate just a few inches above grade and seal the well with a concrete slab, or terminate several feet beneath the surface of the ground and seal the top with a precast concrete slab.

Either above-grade termination can be operated by a hand-operated suction pump, if the well is no more than about 20 feet deep, or a power pump. An underground termination requires a power pump. You can use a suction-type pump here too, but remember the limitations imposed by altitude and the fact that there is always vacuum and friction loss in any pipeline. Those problems can be avoided by installing a pressure pump.

To keep pump noise out of the house and make the pump more readily accessible for service, you can terminate the well in a well house. This can be any small building large enough to hold the equipment and permit ready access to it and plenty of working space to service and replace parts.

Frost can be a serious problem in a well house. When water freezes, it increases by 11 percent in volume. Not only low temperature, but how low, for how long, and the heat volume of the water are important factors. Insulation will defer freezing depending upon these factors, but will not prevent it.

The amount of insulation needed depends upon the microclimate, the amount of heat generated within the well house, and the kind of insulation being installed. What is right in one installation can easily be entirely wrong in another, even if nearby. You can make a reasonably accurate determination by working the same heat-loss formulas used in assessing heat loss for a house or any other building. An easier method, although not foolproof, is to check with your neighbors and area builders to find out what insulating arrangements have been effective in your locale.

Unless temperatures seldom fall below freezing in your area, it's a good idea to provide some permanent heat in the well house, or at least emergency heat, preferably automatic. You can build the well house against the main house, or any heated building, and cut a passageway between the two, providing heat and ventilation and guarding against freeze-ups. You can also wrap the pump and piping with heat tape, preferably the self-regulating type, and connect them through one or more thermostats or thermostatic outlets set to close at 40°F. The tape should be overwrapped with insulation.

Another solution is to terminate the well safely below the frost line. This entails constructing a concrete enclosure with a heavy dirt-tight and waterproof cover (FIG. 15-4). The piping must be trenched below the frost line into the house.

CONSTRUCTION

Get the stave planks ready by chamfering one end. At the opposite end, lop the corners off and bind the end with half a dozen turns of heavy, soft iron mechanic's wire (FIG. 15-5). The chamfered end will drive more easily, and the wire wrap will reduce splitting and splintering when the end is struck with a maul. About 2 inches from the wire wrap bore a 2-inch hole through each plank. This will help when the time comes to pull the planks out of the excavation. If you do not have steel brace rings, you can have them made by a structural steel shop, or you can make your own of wood by cutting planks and bolting the sections together as shown in FIG. 15-6. You can use 12d nails to hold them together, but 16d clinched over is a better arrangement, and 5⁄16-inch bolts are better yet.

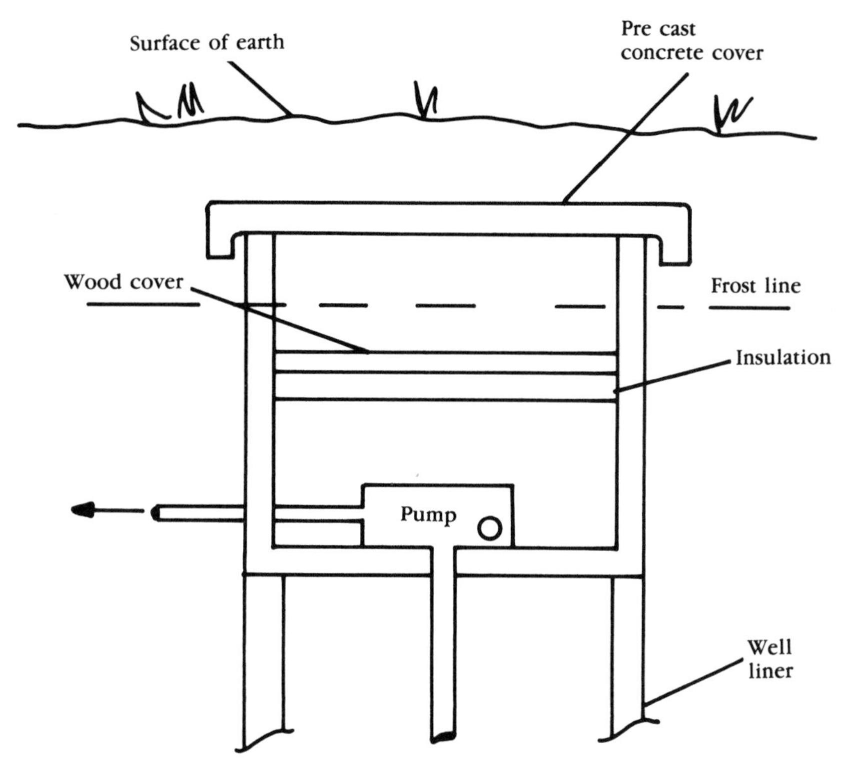

15-4 Here the well pump is housed beneath the soil in a concrete structure and insulated from frost.

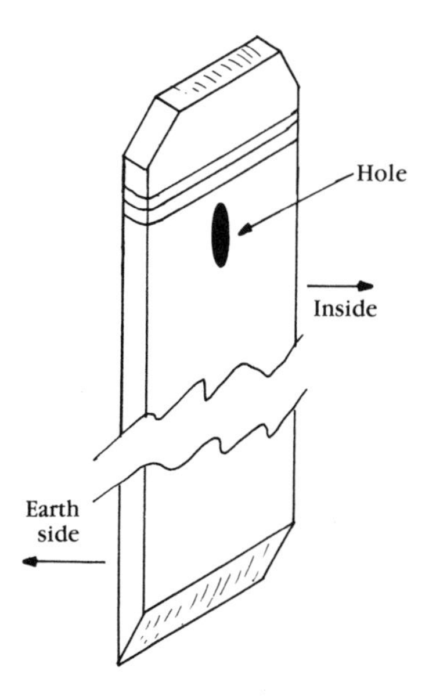

15-5 A shoring plank that is to be driven has its top wrapped with wire to minimize splitting. The hole makes pulling it easier, and the angle cut on the bottom makes for easier driving and tends to force the plank end back against the soil.

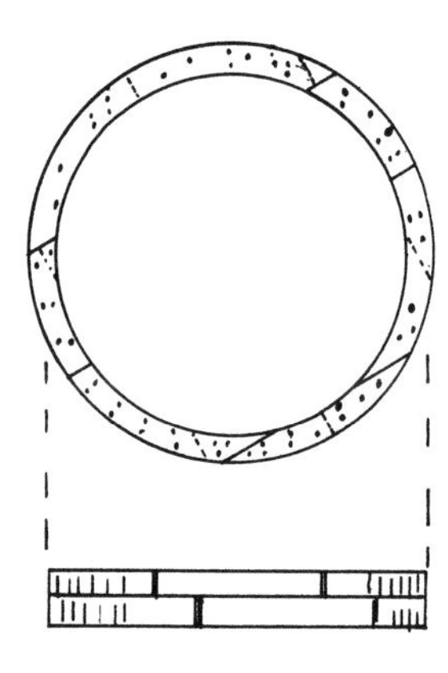

15-6 A retaining ring made by joining a number of curved sections cut from planks.

Excavating

To start the excavation, lay out a circle of the desired well diameter and dig down about 3 to 4 feet all around. Keep the sides vertical and as smooth as possible and the bottom of the hole reasonably level. A gauge made up of 2 by 4 scraps will help keep you properly lined up.

Next dig a narrow perimeter trench all around the floor of the excavation. The deeper you can go, the easier it will be to start the staves. Be sure to leave the center portion that you are working from at the 3- to 4-foot level—that's your safety net and insurance policy, in case a wall collapses.

Position the staves vertically and edge to edge, with the chamfered end down and the chamfer facing inward (FIG. 15-7) so the thrust tendency is outward against the earth wall. Use wire loops or nails just partly driven to keep the planks not being driven from sliding around or falling over. You could even shovel some loose dirt in against the bottom ends to help hold them. Drive the staves into the ground one at a time, a few inches at a time in round-robin fashion, keeping them vertical. Work from a scaffold platform as necessary.

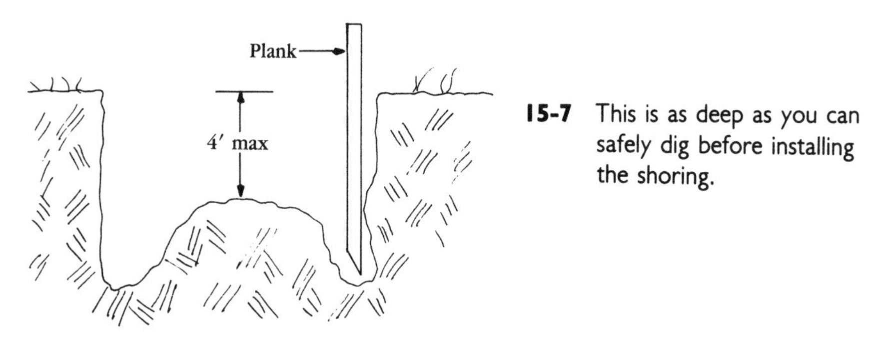

15-7 This is as deep as you can safely dig before installing the shoring.

When the shoring is firmly in place you can remove the dirt island and extend the perimeter trench downward. When you get close to the chamfer on the planks, stop digging and drive again. When you have worked your way down about 5 or 6 feet below grade using this procedure, install the first circular brace. Do not nail it to the staves, but set a ring of 16d or 20d nails around the shoring, all at the same level, and rest the brace ring upon them. Otherwise you'll have a problem trying to drive the staves. The planks tend to move inward, and they will hold the brace secure. Also, once you get beyond about 5 feet deep, keep a ladder propped in place at all times so you can climb out immediately and easily.

Working in water

When you reach the aquifer, water will seep or rush into the excavation. Stop digging around the edges and cut a hole about 1 foot deep or deeper in the center. Place the end of the suction hose to the pump in this hole (FIG. 15-8). Incoming water will run to the hole and the working floor of the excavation will stay relatively dry and firm as the pump clears the water out. Then continue digging at the edges.

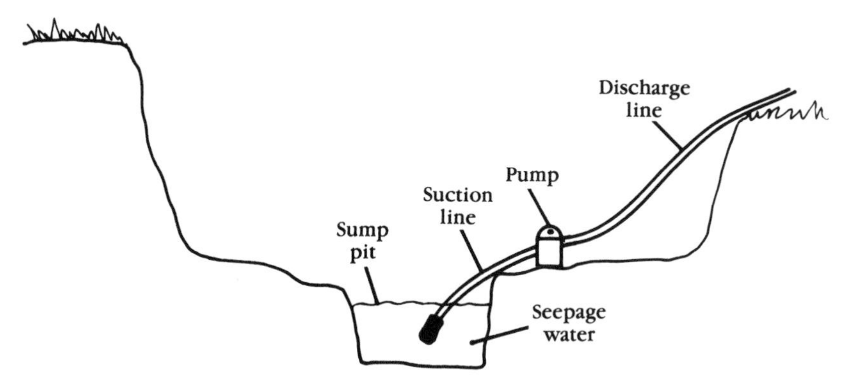

15-8 Dig a sump hole in the center of the pit to catch water seeping in, and pump it out from there.

Constructing the liner base

The depth of the well should include at least 1 extra foot to accommodate the layer of fine gravel you will lay down that acts as a filter to keep sand and dirt from rising into the well. When you reach that point, if you are working in gravel or very coarse sand you can probably lay your first course of block, brick, stone or the bottom edge of a pipe or ring section right on the leveled surface.

If you are in clay or silt or a mixture, or if you have any doubts about the stability of the soil, build a base or foundation for the well liner (FIG. 15-9). One simple method is to lay down a ring of flat fieldstones or cut stones at least 3 inches thick and about 1 foot square or larger. Place them tight together, but use no mortar. You can also pour a concrete base: make an inner form from 1 by 6 boards, plywood strips, or metal curved into a circle about 2 feet less than the diameter of the excavation. Or use a series of short boards in a segmented effect—the form does not have to be round. The wall of the excavation will serve as the outer form.

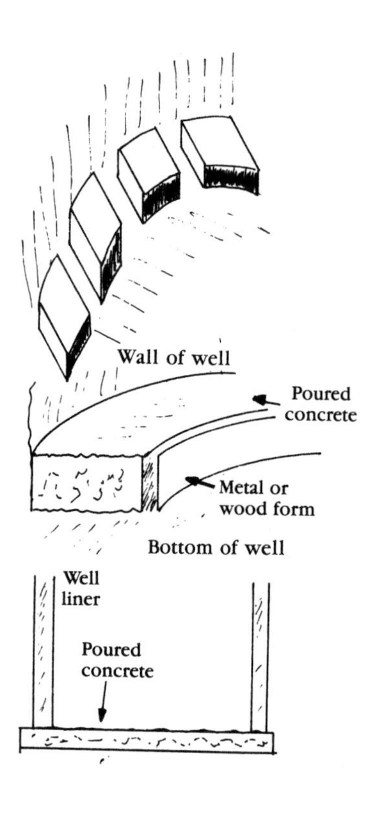

15-9 There are several ways to provide support for a well liner when the soil is too soft. A is a circle of flat rocks, B is a poured concrete ring, and C a poured concrete pad or slab.

You may be working in a couple of feet of water despite the best efforts of your pump. Weight the form down with rocks if necessary, and mix up a batch of concrete that is thicker than normal. Pour it right into the underwater form, taking care not to mix it around much. The concrete will displace the water in the form and will cure normally even though covered with water.

The liner

The lower portion of the liner, within the aquifer, must be pierced in one way or another to allow water to enter. This area is equivalent to the well screen in a piped well, and yield is directly related to its height. Double that and you double the amount of water entering the well, assuming that the pierced area is entirely within the aquifer. The upper portion of the liner must be watertight. You want aquifer water to enter through the screen, and no seepage from soil above the aquifer and from ground level.

When using bricks, blocks or stone, which are always laid flat, knock the corners off to leave openings. Lay up the masonry with mortar consisting of about 1 part portland cement to 2 to 3 parts sand to ¼ part slacked lime or use a mortar cement already containing lime.

Bricks or blocks are laid up in the usual manner. However, as you work, be sure to poke a stick or a trowel between the lopped-off corners of the units to keep the openings clear of mortar. Large stones can be laid up dry; leave a little open space between adjoining stones. From the screen section upward, however, everything is laid up solidly with mortar.

The outside of the liner above the screen section, facing the excavation wall, should be given a ½-inch thick coating of cement plaster. You can use the same mortar mix just described, but add a waterproofing agent.

Packing

Packing a dug well is important. When you have built the liner up to a point about 1 foot above the pierced section, stop. Fill the space between the outside of the liner and the excavation wall with packing material. Ideally, this should consist of graduated gravel or crushed rock along with coarse sand, positioned in three vertical layers (FIG. 15-10). The layer next to the liner should be of stones about 1 inch across. They should be just large enough that they cannot slip through the holes in the liner. The middle layer should be of stones about ½ inch across and the outer layer (against the excavation wall) should consist of coarse sand. No harm is done if the materials become somewhat mixed; don't waste time striving for perfection. You can use a wide sheet of tin bent to a slight curve to keep the materials separated while you are placing them.

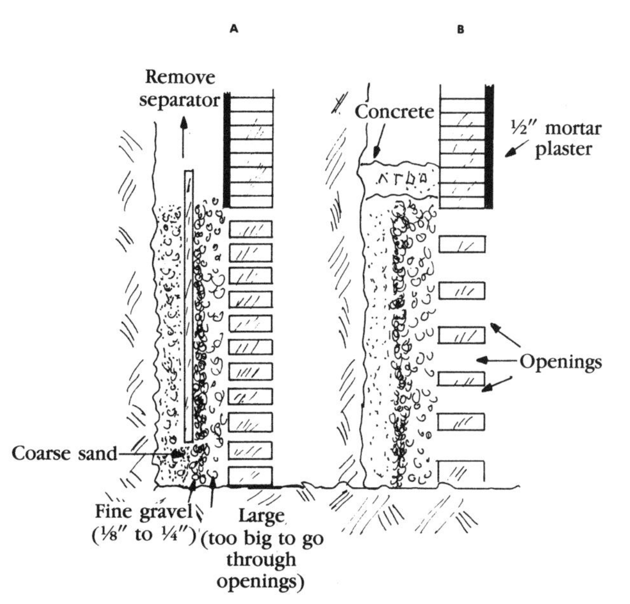

15-10 Artificially developing (packing) a dug well. A. The temporary boards make it possible to place relatively coarse gravel next to the well and coarse sand between the gravel and the aquifer. B. Separators have been removed and the top of the packing has been sealed off.

After the packing and removing separators, pour a seal of concrete about 4 to 6 inches thick over the top of the packing bed, all around the liner. This will keep loose soil from working its way down into the packing and clogging it up or getting into the well. Then go ahead and complete the solid part of the liner.

ALTERNATIVE TECHNIQUE

The technique just described is generally called the *loose stave method* of constructing a dug well. An alternative method is to lock the staves together and sink them simultaneously by undermining the ring in continuous rounds (FIG. 15-11). Variations are to sink concrete rings or pipe sections, a steel ring ("shoe") upon which a masonry liner can be built up as the shoe is sunk, or even a cylindrical steel caisson.

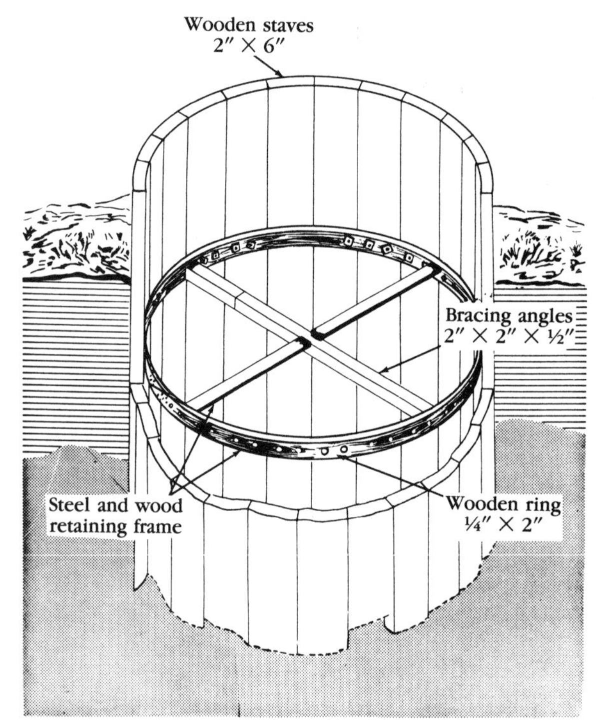

15-11 This method of preventing the collapse of the walls of a dug well consists of building the entire bracing structure above ground and then lowering it as you dig.

The diameter of concrete pipe should be great enough to allow plenty of pick-and-shovel working room inside it. The first one or two sections of pipe, depending upon the height of the sections and the nature of the aquifer, must be pierced. Holes about 1 or 1½ inches in diameter on 12-inch centers in rows 12 inches apart makes an ideal arrangement.

Start by digging a clearance hole about 2 feet larger than the liner outside diameter and about 3 to 4 feet deep. Lower the first section into the excavation, centered and approximately level. Then remove the dirt from the center section of the hole, to a level a few inches or so below the bottom of the pipe. Next work your way around the perimeter, cutting the earth away from under the pipe and 6 or 8 inches beyond its outside wall (FIG. 15-12). As you do so the pipe will drop, and with luck it will not cock and you will not run into boulders. When the upper edge of the first section is 2 feet or so below grade, set and seal the second section. Continue digging this way until you have reached the desired depth.

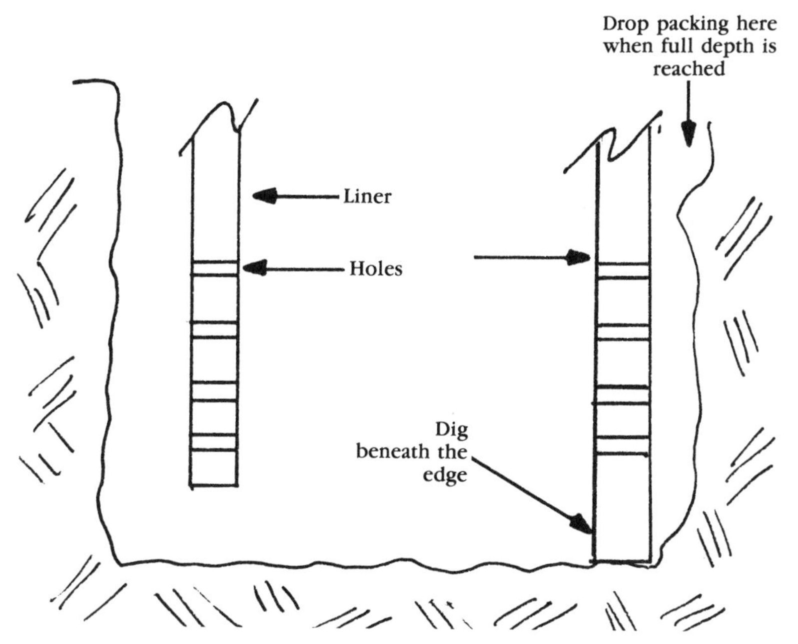

15-12 Dig beneath the edges of the well liner to continuously lower the liner as the well is dug deeper.

To pack this type of dug well, pour coarse gravel or crushed rock—too large to pass through the holes in the screen section of the liner—down into the space between the pipe and the excavation wall until it is filled to about 1 foot above the perforations. You can cover the top of the packing layer with strips of plastic sheeting to keep fine dirt from sifting down into it.

Other types of liners, whether built-up or precast, can be sunk in much the same way.

COMPLETING THE WELL

If the bottom of your well lies in water-bearing sand or gravel you can leave it open. If the bottom is silt or clay, sealing it off is advisable. Do this by slowly pouring in a layer of fine gravel about 6 to 12 inches thick, or about 4 inches of concrete. The purpose of the seal is to keep water movement from stirring up fine sand and silt.

If you have packed the well and topped the packing with concrete, all you need to do is fill the space between the liner and the excavation with clean fill dirt, after removing any shoring materials. If you have built the well around a monolithic liner, pack the lower section of the annular space with coarse gravel that will not pass through the holes in the pierced section. Top this with a layer of fine stone. Then pour a 4-inch layer of concrete. Once the concrete has begun to set up firm—a matter of just a few hours—you can backfill the remaining space. Wet the soil down to make it compact more easily, and leave a small quantity on hand to fill the space left when, with time, the soil subsides.

Chapter 16

Drilled wells

The two main water well drilling rigs are called *percussion cable tool* and *rotary drilling*. Percussion drills operate by means of a bit attached to a wire cable that is raised and dropped repeatedly to open the bore hole that will later be cased to become the well. Debris is removed periodically with a *bailer*, a bucket-like affair. Rotary drilling rigs use a bit fastened to the end of a drill pipe or stem to bore into the earth. Drilling debris is continuously driven from the bore by air, water, or mud under pressure.

Cable-tool drilling, also known as churn-drilling, percussion drilling, spudding, banging and other terms, is much the older of the two basic methods. It is also the simpler method: the rig can be operated by one man when speed is not important. Rotary drilling usually requires three men for safety and efficiency.

Cable-tool drilling is usually the slower of the two methods. The work has to be frequently stopped, the bit withdrawn and the bore cleaned out with the bailer. But in some formations this is the better method. The rigs also cost less, which sometimes (but not always) translates into lower per-foot drilling charges. In addition, lower rig weight and smaller size makes them easier to site in rough terrain. Another advantage is that drilling proceeds with no need for water, drilling mud or high-pressure air.

Some drilling methods combine techniques. Well drilling of this sort cannot be considered a do-it-yourself project, but some knowledge of the systems might help you decide what might work best in your own case and enable you to intelligently discuss the matter.

CABLE-TOOL DRILLING

Cable-tool drilling equipment varies from simple, home-made rigs capable of penetrating only a few dozen feet into the earth, to giant rigs that are mobile and self-contained. The basic principles of operation are identical, and except for size and greater control their parts are very similar.

Equipment

A platform or base, which can, but need not be truck-mounted, supports a ***derrick*** or ***mast*** that is raised from a horizontal traveling position to a vertical working position. The ***drawworks*** includes the engine or motor and the clutch, gears and controls that operate the hoisting line (drawing pipe to the mast), and the ***spudding*** equipment. This is an adjustable eccentric drive that alternately lifts and releases the cable holding the tools.

The tool cable runs from a reel up to a ***sheave*** (pulley) at the top of the mast and then down to the tools, which together are called the ***string*** (FIG. 16-1). The cable is fastened to a socket, which in turn is attached to a ***drilling jar***, which is screwed fast to a ***drill stem*** to which the ***bit*** is secured.

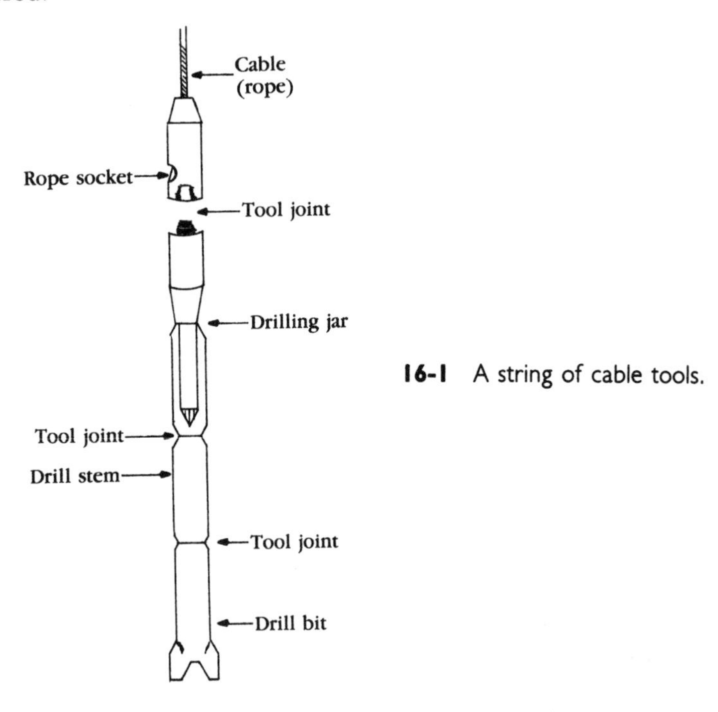

16-1 A string of cable tools.

The jar can best be described as a section with a joint so loose that its two halves can separate by several inches. When the cable is lifted, there is a hammer action upward against the top of the jar.

The bit is the business end of the string. The cutting edge of the bit flares out beyond the diameter of the bit's body so it will bore a hole larger than the diameter of the tool string following. For water well work, the bit is several feet long, as is each part of the string. The weight of the string provides the force that drives the bit. And the weight of the jarring tool provides the impact that frees the bit when the spudding drive lifts the cable and pulls the whole string upward.

The bailer is essentially a long pipe fitted with a check valve at the lower end. When the bailer is used the tool string is withdrawn and the bailer attached to a second line called a *sand line* or a *bail line*, that also runs through a sheave at the masthead. The bailer and bail line are dropped into the bore hole and the weight drives the bottom end of the bailer into the cutting debris at the bottom of the bore, forcing the check valve open and filling the bailer. As the bailer is retrieved the check valve closes and locks in the debris. A *sand pump* is a bailer with a rod and plunger inside it that performs the same function. This provides a suction that pulls debris into the end of the pipe.

Bailing continues until the bottom of the well is clear and clean. The cutting debris is dumped near the well, then later disposed of. Frequent bailing is important because the accumulated debris at the bottom of the bore cushions the impact of the falling drill bit and considerably reduces its effectiveness.

Drilling

Cable-tool drilling is always done with what the drillers call a "tight line." The release of the cable should be so timed that the cable is never slack, as it would be if allowed to simply unwind off the reel faster than the travel of the bit down the bore hole.

When the cable follows the tool, it always twists somewhat, causing the bit to rotate and strike a slightly different spot with each stroke. A taut cable saves time because there is no need to spool up slack. As soon as the jar reaches the limit of its travel the bit is immediately pulled upward.

The second or two that might be wasted pulling up cable slack might not seem important, but it is. *Drilling speed*, the rate at which the string is raised and lowered, can be as high as 60 strokes a minute. A second or two lost on each stroke cuts the speed to ribbons. The operator must set the rate, and must adjust the spudding drive to the nature of the formation being drilled and the action of the cable and tool string. This is done by varying the speed and the length of the stroke.

The bit must be gauged frequently. If its cutting diameter wears beyond a quarter inch the bit will "freeze" in the bottom of the bore hole. Bits can be sharpened in a portable forge right at the drilling site. Some drillers retain worn bits for use farther down the bore hole.

As well depth increases, pipe diameter will be decreased. When a casing cannot be driven any deeper, the only way to proceed is to slip a smaller diameter casing into the in-place casing and continue down. Wells that are expected to go thousands of yards deep will start off with casing 36 inches or more in diameter, with a tool string to match. Such super depths are only achieved with rotary rigs; percussion drilling is generally limited to several hundred feet maximum.

MUD-SCOW DRILLING

Mud-scow drilling (FIG. 16-2) depends upon a large, heavy bailer to produce a relatively large diameter hole in sand, clay, mixtures of quicksand and clay, or comparatively thin layers of quicksand and clay. The bailer is similar in design to the one described above. It has a greater size and weight, and a *drive shoe* is attached to the bottom end. This is a sort of coupling or collar screwed or welded fast that has a knife edge for quicker, easier driving.

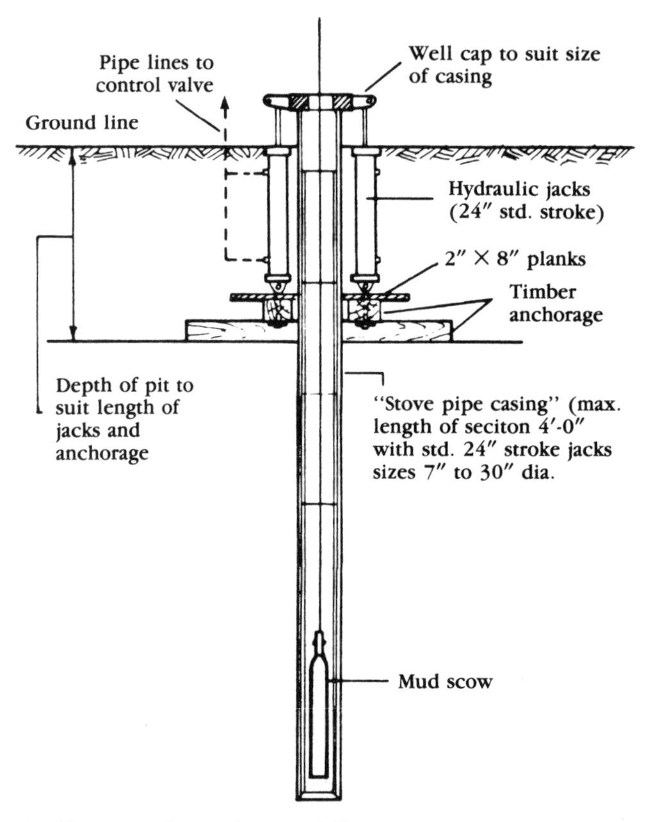

16-2 Elements of a mud-scow drilling operation. Courtesy of Koehler, Inc.

Mud-scow bailers are connected to jarring tools that permit longer jarring strokes. Stroke lengths for the string are typically 24 to 36 inches, considerably shorter than for ordinary bailers. Mud scows are run at average working speeds of 25 to 40 strokes per minute, a relatively slow pace. Stroke length depends upon the formation being drilled. In sand and gravel the stroke is kept short, while in clay it can be much longer.

When percussion drilling drives through stable, noncollapsing formations the casing can be lowered into place after the bore has been completed. In collapsing formations the casing must follow directly behind the bit. The end of the casing always precedes the working end of the scow by a short distance. Should the scow get ahead of the casing, the scow will probably jam. The whole purpose, and the basic design, of the scow is to drill into caving soil—not easily accomplished with standard percussion techniques and equipment.

Another major difference between mud-scow and standard percussion techniques is that the mud-scow casing is "pulled" into the formation, not driven.

Work starts with the excavation of a flat-bottomed hole roughly 10 by 10 feet square or larger and 6 feet or more deep, centered on the bore hole. A shallow bore hole is then dug by hand. Next 3-by-8-inch timbers are laid down to form an anchor or skid to which double-acting hydraulic jacks are bolted in a vertical position. The excavation is then filled with crushed rock or other fill to hold the timbers in place. The casing is positioned vertically and a suitable cap placed on top.

The jacks exert a downward pull on the casing, driving it a distance into the earth. Now the mud scow is put to work inside the casing. As the scow is lifted, dropped, and removed for emptying, the jacks continue their downward pull on the casing.

Because the work is in collapsing soil, the walls of the bore tend to collapse inward as the scow removes the soil. The pressure of the hydraulic jacks forces the casing to advance downward into the space partially cleared by the scow and partially produced by the collapsing formation.

JET DRILLING

Commercial jetting uses the same principles I discussed in Chapter 14, but the size and power of the equipment is greater, and percussion is used to drive the well casing.

This technique is generally limited to 2- to 4-inch diameter wells (inside diameter at the top of the well casing) that seldom reach more than 100 feet. The equipment is relatively light and simple, consisting of mast and drawworks plus a high-power, high-pressure water pump, all usually mounted on a light truck.

A hollow bit perforated at the sides near the cutting edge is used. The bit is screwed fast to a drilling pipe and water is forced under high pressure down the pipe and out through the bit holes. The casing is usually first driven down a few feet, then the bit follows into the casing and is lifted and dropped with short, fast strokes.

The drill bit chops up and otherwise loosens the formation. The high-speed flow of water from the bit holes washes the cuttings and loose soil up the annulus—the space between the drill pipe and the casing wall—and out the bore hole. In sand, soft earth, gravel, and similar loose formations, this method of drilling is very fast. The casing either drops of its own weight or is driven to speed its descent

Hollow-rod drilling

Hollow-rod drilling is a variation of percussion-jet drilling. No water pump is needed, and only a small amount of water. The system is most effective and efficient in comparatively soft formations like clay and sand, and is seldom used for wells over 4 inches in diameter.

Equipment consists of the usual mast, drawworks, spudding equipment, and associated items. Drilling is done by means of a special bit at the end of the drill pipe that contains openings and passages leading upward. The bit is mounted within a short length of tubing that has a cutting edge at the lower end. The tube is screwed into a check valve, which is in turn attached to the drill pipe.

The bit and drill pipe are set vertically into a shallow prepared hole then repeatedly raised and dropped. The hole is kept filled with water. The bit cuts the formation, and the cuttings and loosened soil mix with the water and form a slurry. Each time the bit is dropped, some of the slurry is forced through the bit openings into the bit, past the check valve, and into the hollow drill pipe. When the pipe is partly filled the string is removed from the bore hole, tipped over and emptied, and the process is repeated.

ROTARY DRILLING

Rotary drilling is similar to the power boring methods described earlier, and also to jet drilling. Figure 16-3 shows the layout of a typical rig. The working end of the drill is a hollow, perforated bit attached to a hollow drill pipe. This is fastened to a shaped section, the kelly, that passes through a matching hole in a flat, round, horizontal table. The drive engine rotates the whole assembly, which is suspended vertically from a mast or derrick and handled by the drawworks. As the bit turns it chews into the formation and the cuttings are brought to the surface by a continuous flow of drilling mud that is pumped down the drill pipe and out through the bit perforations. The mud emerging from the annulus is routed to a settling pit where the particles of stone and sand settle out. The mud is filtered, withdrawn, and pumped back down the drill pipe in a continuous operation (FIG. 16-4).

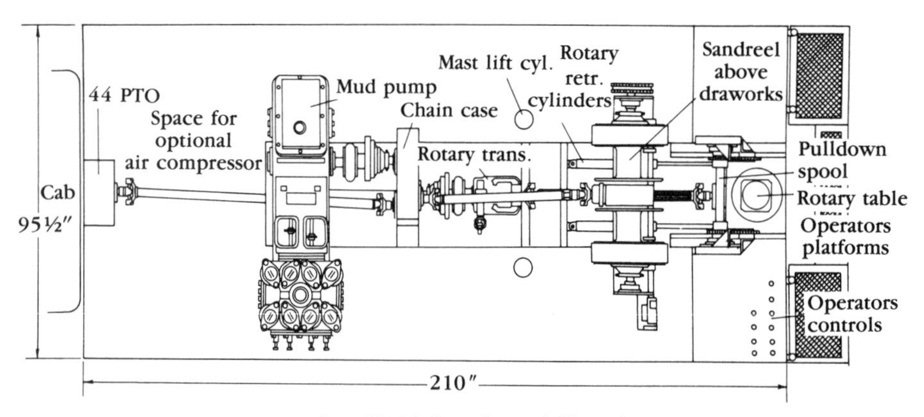

16-3 Principal parts of an SS-10 Speedstar drilling rig. Courtesy of Koehler, Inc.

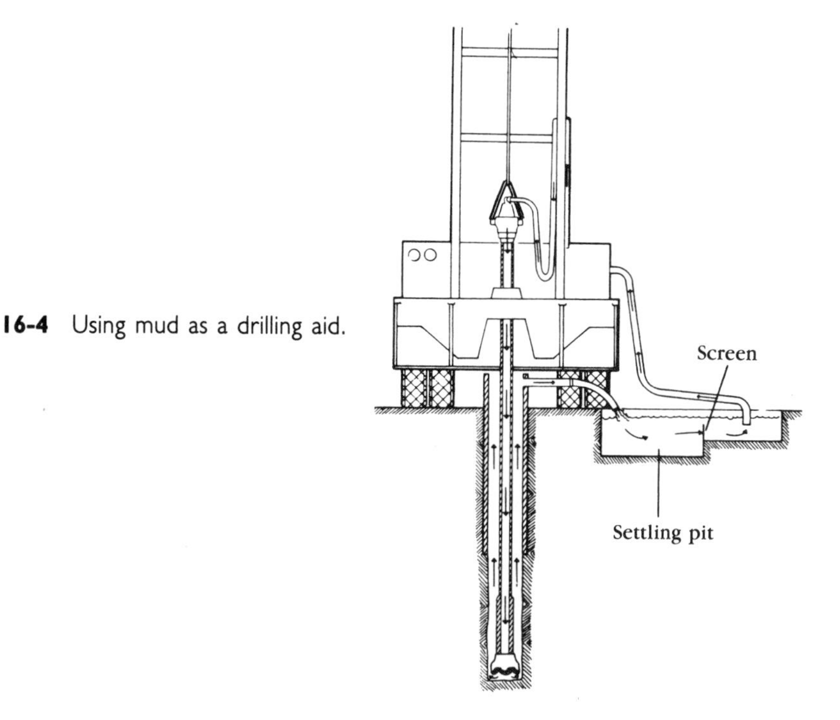

16-4 Using mud as a drilling aid.

Drill bit design varies with the type of formation encountered. Simple fishtail or star-shaped bits are used in soft formations. Bits with short tooth cutters are used in hard formations. In medium-hard formations, cone or roller bits work best. If the formation is very hard, bits fitted with carbide buttons are used.

Soft-formation bits are usually rotated at a speed of 50 to 150 rpm. Speed is reduced to 30 to 50 rpm in hard formations. The pressure on the bit is produced by the weight of the tool string. If that is insufficient, extra-heavy sections of drill pipe are added. Weight varies from 2000 to 5000 pounds per square inch of bit diameter, depending upon the hardness of the formation; the harder the formation, the greater is the required pressure.

Large-diameter holes

Bore holes with a diameter of more than a foot or so are usually drilled in stages. A small-diameter hole, a pilot hole, is drilled first, then enlarged with a roller bit.

When a hole directly beneath a casing must be enlarged to enable the casing to drop or be driven down, one of two types of *under reaming* tools are used. One depends upon hydraulic power. A fluid under high pressure is jetted through holes in the tool and against the walls of the bore hole. The other type of tool has pivoting blades that spring outward when hydraulic pressure is applied.

The mud

Sometimes just called mud, and sometimes drilling mud or drilling fluid, mud is not just dirt and water. It is a mixture of various chemicals and mineral substances such as bentonite, a colloidal clay.

Mud performs a number of important functions: it removes the cutting debris; it cools the bit; and it prevents the bore hole walls from collapsing upon the drill string. But the mud can also create problems at the lower end of the bore hole, where the well screen will eventually be positioned.

The turning action of the drill string and the pressure on the mud produces a firm layer or smear of clay on the wall of the bore. The mud has to be controlled so that proper sealing can be accomplished without penetrating the walls to such a degree that developing the wall later on is difficult. The mud seals the well wall and stops or slows the entrance of water until it is removed during the developing process.

Experienced drillers maintain that mud control is often the most important aspect of rotary drilling. Mud balance, meaning the weight per gallon of the drilling fluid, is closely watched and generally held to 9 pounds per gallon. The pH of the mud is held between 8.0 and 9.0. Sand content is kept below 2 or 3 percent.

AIR-ROTARY DRILLING

Air-rotary drilling employs the usual mast and drawworks, but the motor or engine does not directly drive the rotary table and kelly to turn the drill pipe. Instead, it drives an air compressor which in turn drives the table. This arrangement is further classified as an "air-rotary table-driven" rig.

When the drill string is not rotated by a table and kelly but instead by a slide-mounted motor secured to the mast atop the string, it is called a "top-headed drive" rig. The slide arrangement permits the drive motor to remain attached to the mast and still power the descending drill string.

A variation on both of these types consists of the addition of a hydraulic pump and the replacement of the air-drive motors with hydraulic motors.

Further application of air power

Whether the drill string is rotated by air or hydraulic motors, all air-rotary drilling uses a constant stream of high-pressure air driven down the drill string and out through holes in the bit. The rapidly moving air clears the face and cutting surface of the formation and carries the debris up the annulus to ground level for disposal.

Neither mud nor water are needed, so no layer of water-sealing clay forms on the surface of the bore hole. A disadvantage is that the system can only be used in noncollapsing formations, or in collapsing formations with the simultaneous insertion of a casing.

Roller-type bits are usually used when drilling hard formations, with shorter teeth for very hard rock and longer ones for softer formations. Rotating speed ranges from 10 to 20 rpm in very hard rock, and can be increased as hardness decreases. Downward pressure is also adjusted to the relative hardness being encountered: the softer the rock, the less pressure is needed.

AIR-HAMMER DRILLING

Air-hammer drilling is a modification of air-rotary drilling. The rotating hollow drill pipe with attached bit is employed as usual, but a special drill bit is continuously pounded into the rock by an integral jackhammer arrangement. This speeds the cutting action considerably. The rotating action lets the bit strike in a slightly different spot with each blow, preventing it from locking up in the formation. This also keeps the bore vertical, while the weight of the tool string adds force to the cutting edge of the bit.

The air hammer attached to the bottom of the drill pipe consists of a piston, one or more valves and the cylinder in which the piston moves. Designs vary, but they all operate the same way. Air under pressure comes down the drill pipe, enters the cylinder through a valve, and drives the piston down. The piston strikes the top of the bit, driving it down. A valve redirects the air flow to the underside of the piston. The piston is raised, making it ready for the next cycle. Expelled air is directed towards the face of the cutting edge of the bit. The air scours the bottom of the bore hole and carries the debris up the annulus and out of the bore (FIG. 16-5).

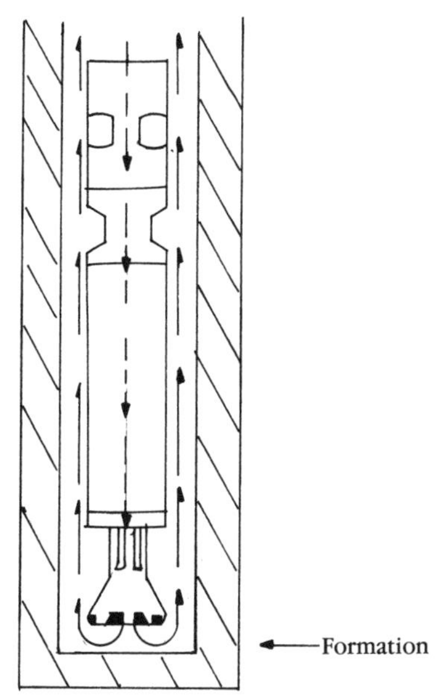

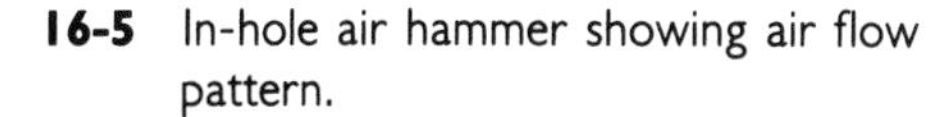

16-5 In-hole air hammer showing air flow pattern.

To facilitate changes and refacing, all air-hammer bits are interchangeable. The driller can remove a bit to shape and possibly retemper it without losing drilling time. Bit faces or cutting surfaces are not alike. Some have two or more air holes in the face, blowing exhaust air directly across the cutting edges and onto the surface being drilled. In other designs the air holes are at the sides of the bit.

Air-hammer drills are usually rotated at speeds of 5 to 35 rpm: the harder the rock formation, the lower the speed. The reciprocation rate of the bit itself is typically 60 cycles per minute. The exact details vary with down-hole air hammer design, air pressure, bit size, and the characteristics of the formation. The harder the rock, the faster the bit rebounds.

Typical air pressure requirements range from 125 to 250 psi, much higher than the 30 to 50 psi needed in straight air-rotary drilling. Depth limitations vary according to equipment and conditions, but 1000 feet is typical. Air volume, air velocity, bore hole diameter, and drill pipe diameter are all interrelated. Compressor capacity must match needs.

REVERSE-CIRCULATION DRILLING

The principal parts of a reverse-circulation drilling rig are shown in FIG. 16-6. This method is considered by some drillers to be a form of excavation. The technique is best suited to sand and gravel formations and is a good choice for collapsing soils, but is not suited to any formation harder than soft rock. In sand and gravel it is the fastest way to open a large-diameter well hole. Diameters currently feasible range from a minimum of 10 inches to a maximum of 60 inches, sometimes more. A depth of 1500 feet with a 10-inch bore is sometimes possible. In the larger diameters, as much as 600 feet of depth is practical.

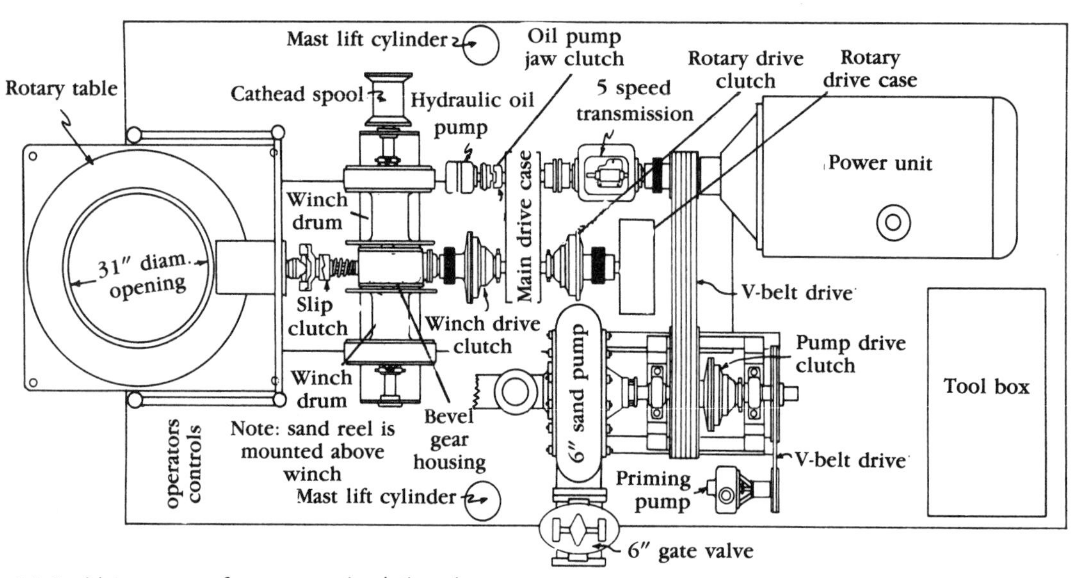

16-6 Major parts of a reverse circulation rig. Courtesy of Koehler, Inc.

In other rotary-drilling systems water, mud or air is driven down the drill pipe to force the drilling debris up and out of the annulus. In reverse-circulation drilling water is pumped down the annulus and sucked back up the drill pipe along with the cutting debris. The bore walls never become sealed with a mud paste that inhibits water flow and later must be removed.

The quantity of water required is considerable—roughly three times the volume of the planned hole. This water is collected in a settling pit, where the solids drop to the bottom and the clear water is pumped back down the bore hole in continuous recirculation. On an average well, water is circulated at a rate of about 1000 gpm (gallons per minute).

The speed of the water does not affect the walls of the bore. Because the annulus volume is considerably larger than the drill pipe volume, water traveling down through the annulus moves at roughly 1/20 that of the water ascending in the drill pipe.

The addition of compressed air to the water drive arrangement results in an even more effective drilling rate. Increased fluid velocity results in faster scouring at the bottom of the bore, faster clearance of blockages caused by collapsing soil, and deeper well penetration. Rates averaging 40 feet per hour have been achieved.

Chapter 17

Well screens

A well screen is sometimes called the "business end" of a well because of its importance in the efficient performance of the well. The screen must prevent unconsolidated material, like sand, from entering the well pipe, reaching and damaging the pump, and interfering with the use of the water. At the same time, it must not interfere with the water flow, which would reduce head pressure. The screen must also act as a section of casing.

There are many companies manufacturing well screens to industry-adopted standards. These screens are more expensive than homemade alternatives like slotted pipe, but the difference in performance and service longevity more than makes up for the difference in initial cost. Properly designed and constructed well screens have continuous, uninterrupted slots circumscribing the screen's circumference. There is a minimum of solid space between the slots, leaving a maximum of open space for the passage of water. The slots have a vee shape so that the smallest opening is at the exterior surface of the slot, increasing toward the center of the screen.

Only a single metal is used in screen construction, virtually eliminating galvanic corrosion between the screen and a well pipe made of a different metal. The screen must be strong enough to resist deformation during handling and placing and to resist the pressure of potential bore-hole wall collapse. The screen should also be provided with an end fitting to make the screen adaptable to a wide variety of installation and operating conditions.

The best screen design appears to be the continuous-slot screen made by wrapping cold-drawn wire with a triangular cross section around a number of rods arranged to form a cylinder. The triangular wire makes contact with the rods along the knife or thin edge. All these contact points are then joined permanently to the rods by welding. This results in a single, one-piece assembly that is rigid and strong.

The metals most often used are various specialized non-corrosive steel alloys, silicon red, red brass, Monel, types 304 or 316 stainless steel, galvanized Armco iron, or galvanized low-carbon steel. For especially corrosive wastes—which would not be drinking water—special alloys are used. Plastic has recently entered the well screen field; it is strong, long-lived, and resistant to corrosion.

The opening-to-wall area of a well screen is very low. Actually, the openings are no more than a series of horizontal slots usually spaced farther apart than the width of each slot opening. Where the aquifer is generously supplied with water and the need for a high draw rate is low, plastic screens (and plastic well pipe) become practical and economical.

Another, but less desirable, way to construct a well screen involves forcing the wire into slots or notches cut into rods or bars. No welding is required, and the result is a savings in manufacturing costs at the expense of screen rigidity and strength.

The continuous-slot, welded-in-place well screen can be readily manufactured with almost any size of opening or spacing between turns of wire. If a slot opening of 0.090 is called for, the successive turns are spaced that far apart. If a narrower opening is preferred, like 0.020, the rolling and welding equipment can be set to that figure.

If there is a need for a change in openings between the lower and upper ends of the screen, or even between the middle and the ends, this too can be accomplished without any difficulty during fabrication. The need for varying slot spacing in a single screen might appear unnecessary, and it is in a screen 5 or even 10 feet long. But some screens in commercial or industrial installations can run to 30 and more feet, with outside diameters of 36 inches.

The value of a triangular cross section of the wire—which, welded in place, results in slot openings that are smaller at the outside surface than they are inside the screen—is that the slot openings do not clog as readily as would parallel-sided openings, or openings that are larger at the outside of the screen than on the inside.

When a slot has a narrow opening, as the sharp edges of these slots do, a grain of sand that passes through the opening easily passes on. But if the slot has parallel sides, a marginal-size grain of sand has to squeeze past the entire length of the slot's walls to get through the screen and travel beyond it. The chance of the grain loading in the slot is therefore much greater. And when a slot opening decreases in size from exterior to interior, it becomes a trap for all grains small enough to enter but too large to pass all the way through.

Developing a well depends upon passing the fine grains of sand and silt through the screen openings. If particles become caught in the screen and clog it, development cannot be satisfactorily accomplished. The continuous slot type of screen has only two possible contact points that can impinge upon the movement of a particle passing through the screen and into the well pipe. For all practical purposes, this screen can be considered as nonclogging.

Well screen efficiency is measured by the ratio of intake area per square foot of screen surface, and water intake velocity. All the wire types of well screens have higher intake area ratios than any other present-day screen designs. Screens with triangular-shaped wire have the lowest water intake velocity. The result is that less head or water pressure is lost with this type of screen with any other.

Other screen designs

Another major screen design is the *louver* type. From the outside the horizontal slots appear to be small louvers. The use of these screens is limited to gravel aquifers or gravel-packed wells.

Still another design consists of pipes that have been slotted vertically. A slotted screen can be commercially manufactured, but more often it is made on the job site with a cutting torch. A continuous-wire screen will on the average have 8 to 10 times the open area that a slotted screen of equal diameter has. Government tests have demonstrated that a 10-foot length of 8-inch commercial continuous-wire screen performed four to six times better than a 20-foot length of 12-inch home-slotted pipe.

Plastic pipe performance or efficiency is also greater than the home-slotted variety because the plastic has proportionally more open area. However, plastic pipe is only ⅙ to ⅒ as strong as stainless steel or a special steel alloy. Therefore, plastic screens are limited to small-diameter wells that are not too deep.

Considerations

Welded, continuous-slot screens are available in two different series, *telescoping* and *pipe-size*. A telescoping screen has a slightly smaller outside diameter than the inside diameter of the well casing in which it will be used. A 4-inch (trade or nominal size) screen, for example, typically has an outside diameter of 3¾ inches, allowing just enough clearance to lower the screen down through a well pipe or casing with a 4-inch internal diameter to its position at the bottom of the pipe. Lowering the screen through the casing is the method most often used because it is the surest and safest. Telescoping screens are usually furnished with lead expansion rings at one end.

Pipe-size screens, also called ID screens, have inside diameters equal to and are designed to mate up with standard pipe sizes. They are used when it is necessary or desirable to maintain the same casing diameter from top to bottom of the well, or where for some other reason the well screen must be permanently attached to the casing or well pipe and lowered with it. Pipe-size screens are made with either welding rings or pipe threads at each end. The threaded connections are rarely used on pipe-size screens with diameters larger than 12 inches.

Slot openings for continuous-slot screens can be manufactured with openings from as little as 0.006, or 6 thousandths, of an inch. A screen with a No. 10 slot, for example, would have openings 0.010 or 10 thousandths of an inch wide. Size is crucial: too large a slot admits too much sand, and too small a slot impedes water flow.

Although pipe-size and telescoping screens are the most popular, there are other styles. One is the louver or shutter type, manufactured with a variety of opening sizes. It generally comes in 5-foot lengths designed for welding in place.

The pipe-base screen is another type. The core of the screen is a heavily perforated steel pipe whose main purpose is to provide overall strength. A continuous-slot screen tube is mounted on the core. In some designs the wire screen is wrapped directly onto the surface of the pipe and welded in place. In another design a number of bars or rods are spaced equally apart and welded to the pipe core. The wire screen is then wrapped around the bars and welded to them.

Because the bars hold the wire a distance from the slots in the core, the second type of screen is more efficient than the first. Nevertheless, in both designs the water must first flow through the spaces in the screen, then through the pipe slots. The result is much greater pressure loss in a pipe-core screen than in the straight screen design.

Still another variation consists of welding a standard, prefabricated continuous-wire screen onto a pierced pipe. This is supposedly stronger than the other two pipe-screen styles, but its efficiency is just as low.

Drive-well screens

Also called *drive points*, drive-well screens can be made by attaching a steel or bronze drive point to one end of a continuous-wire screen and a threaded collar to the other. The maximum diameter of the drive point is just where it screws onto the screen, and is slightly larger than the screen itself. This lets the drive point push stones slightly outward and past the sides of the screen, protecting it.

For further protection a brass jacket well point is often used. This consists of a perforated bronze or brass pipe covered with a wire mesh screen that is in turn covered by a perforated brass sheet. This well point design carries screen protection just about as far as it can go. Another variation eliminates the screen altogether. The core is a perforated or slotted brass pipe covered by a slotted brass tube spaced a distance from the inner pipe.

Whereas continuous-slot wire screen openings are specified in terms of the width in thousandths of an inch, mesh-covered well point openings are designated by the number of openings per lineal inch. Popular sizes are 40, 50, 60, 70, and 80.

ANALYZING THE AQUIFER

The sizes of the grains of sand and stones that comprise the aquifer and their size distribution and mix (if mixed at all) determines the size of the well screen openings you must select. When all the grains or stones are alike in size, there is a physical limit to how tightly they can be packed. When there is a mix, packing can and generally is much closer, because the smaller particles slip into the spaces around the larger ones.

If the aquifer consists of a mixture of fine and coarse sand the permeability of the mix is far less than that of coarse sand alone, but only a little more than fine sand alone. This assumes that there is sufficient fine sand to completely fill the voids around the coarse sand. If the particles in the sand or sand and gravel mix are present in a range of sizes, permeability is not significantly greater than that of the finest grains of the mix. In other words, the voids around particles will not be larger than that which is formed between the smallest particles present.

Analyzing sand samples

Professional well drillers plan and design their wells on the basis of formation analysis. They refer to existing records of water depth, aquifer characteristics, overburden and formation type, then drive a test well. Core samplings are constantly taken and analyzed. The results are used to determine the best screen opening size.

Briefly, analysis for this purpose consists of sifting a measured quantity from the samples through a series of sieves. Sieves with decreasing opening sizes are used for sand and gravel, six for sand alone, and five for fine sand. The first and coarsest is a 6-gauge mesh with 0.131-inch openings. The last and finest is 100-gauge mesh, with openings only 0.006 inch across. In use, the sand-gravel mix is sifted through the first screen. Whatever passes through is sifted through the next smaller size screen, and so on down through them all. The material that is collected by each sieve is weighed and the weight noted. The data is then plotted as a curve on a graph (FIG. 17-1).

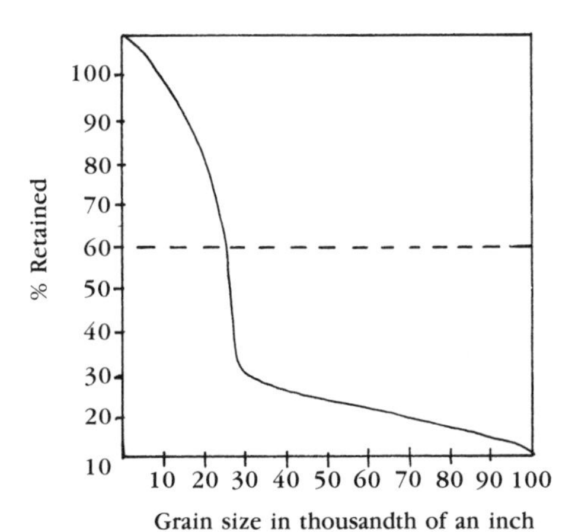

17-1 Typical curve plotted when sand or gravel is passed through a series of screens of diminishing mesh size, to determine the percentage of grains retained by the various sizes. To keep grains larger than 30 thousandths of an inch out of your well, you would choose a screen with similar-sized openings. As you can see from the dotted line, 50 percent of the sand is finer than 30 thousandths of an inch.

The graph data is used to estimate the best screen opening dimensions for the well. The most efficient opening size is that which at least equals in percentage the voids in the aquifer or gravel pack and can at the same time allow construction of a screen strong enough to withstand the weight and pressure of the surrounding aquifer.

Let's assume that the volume of voids, or the porosity, of the surrounding aquifer is 25 percent. If the installed well screen has a porosity of no more than 15 percent, the screen will impede the flow of incoming water from the aquifer. There will be a greater pressure loss at the screen than is available in the aquifer.

In actual practice, the open area percentage of the screen selected is often greater than the volume percentage of the natural voids in the surrounding aquifer. The reason is that development procedures will increase the porosity of the adjoining aquifer. This removes the silt and fine sand from the aquifer in the vicinity of the screen, thereby increasing the porosity of the surrounding portion of the aquifer.

Selecting the screen

By rule of thumb, the selection range of screen slot opening sizes is that which will retain 40 to 50 percent of the sand in the surrounding aquifer. This range represents roughly the midrange of sand grain sizes. Half the grains will pass through and half will not.

This is not difficult when the sand has been sieved and the sizes graphed, and when the size mix is fairly even—the line on the graph is reasonably straight and the sizes divided about equally. In that case the vertical line on the graph, which represents percentage of retention, is followed to the right until it intersects with the grain-size curve. The point carried down to the required slot size is indicated by the grain size in thousandths of an inch.

Grains larger than the desired slot size, which lie to the right of the size indicated, will not pass through. Smaller grains will. The curve simply indicates how much of the sand is smaller-grained and how much is larger than the chosen percentage.

When the graph curve is not smooth, this indicates that there is considerable abrupt variation in grain sizes. In such cases a compromise slot opening size is selected.

It is desirable to retain less than 50 percent of the surrounding sand when it is believed that development will remove the fine sand from the immediate area around the well screen. Slightly larger slot openings are also selected when the water quality will cause the slot walls to become mineral-crusted; the larger openings offset the reduction of open area caused by the encrustation. And when the aquifer lies immediately below a layer of very fine sand, slightly smaller openings are selected because in time the fine sand will be drawn down through the aquifer and into the well.

Gravel-packed wells

Wells are artificially packed with gravel whenever indications are that the process will be faster, easier, and cheaper than natural development (see Chapter 18). It is also done when natural gravel packing is impossible because the aquifer consists of fine sand and does not have a mixture of grain sizes.

Packing around a screen acts as a primary screen (FIG. 17-2), making it possible to employ larger slots in the well screen than would otherwise be practical. Larger screens mean lower restriction of water flow at the screen, and consequently greater yield.

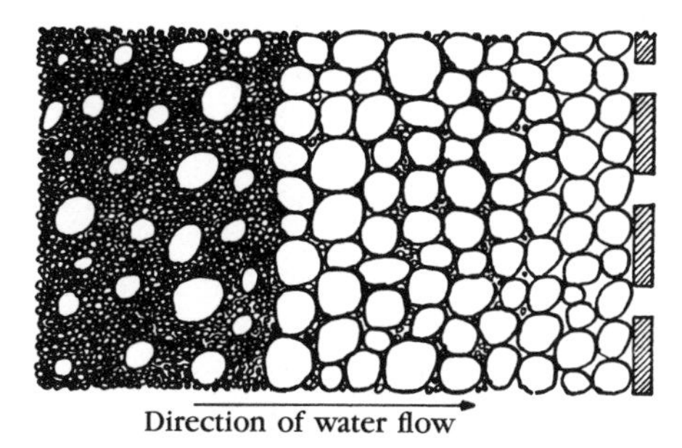

17-2 Fine sand acts as a filter for coarse sand, which in turn acts as a filter for gravel.

Gravel size is determined on the basis of the sand that comprises the water-bearing formation. If the gravel is too coarse, fine sand can work its way into the voids in the gravel pack and so clog the gravel. If the gravel is too fine, the artificial pack could itself reduce water flow. In addition to the care exercised in size selection, the gravel must be screened so all the stones are the same size.

For medium-to-coarse sand formations, gravel up to ¼ inch in size is generally selected. For fine sand, gravel no more than ⅛ inch in diameter is used. Generally, slot and gravel size is selected to permit retention of 75 to 90 percent of the gravel used.

Water velocity

When water is pumped from the well, it is replaced by water flowing in from the aquifer. The velocity of the water flowing inward through the well screen has an important bearing upon the life and the efficiency of the screen. Field and laboratory tests have proved that when this water travels at a velocity of 0.10 feet per second, friction losses caused by the screen slots are minimal, encrustation rate is minimal, and the corrosion rate is low. Slower velocity is even less detrimental.

Water velocity through screen openings can be calculated by formula, but for ordinary single-family domestic water use this is hardly necessary. To quote some relevant figures: a screen 3 inches in diameter and 1 foot long with size 10 slots will pass 3.10 gpm; size 40 slots will pass 9.9 gpm; and size 60 slots will pass 13.0 gpm at a speed of 0.10 feet per minute velocity or less.

At first glance 3 gallons per minute may not seem like much, but that works out to 180 gallons an hour and 4320 gallons every 24 hours. That is a lot of water when the average household seldom needs more than a few hundred gallons of water per day.

Screen diameter

The prime consideration in selecting well screen diameter is whether or not the well pump will be installed within the well casing, most often done in deep wells. An in-well turbine pump is selected because a pump near the bottom of the well is more efficient than one near or on the surface. Only turbine pumps have the physical configuration (small diameter) at the required horsepower.

Well casings are not always the same diameter from top to bottom. On a deep well driven through a caving formation the casing diameter must start large and be reduced with increasing depth. The pump diameter is selected on the basis of the casing diameter at the desired pump depth. In such cases the screen diameter can be less than the casing diameter. In any event, the overall outside diameter of the pump should be such as to provide a minimum of 1 inch of clearance between it and the well casing. By rule of thumb, the pump selected is two sizes smaller than the well-pipe diameter.

When working with a well point screen, the screen diameter is close to that of the associated pipe. For very shallow driven wells, minimum-diameter pipe can be used. When you have to go below 20 feet and the formation offers increasing resistance to penetration, you have to switch to larger diameters of pipe just to gain the necessary stiffness. Screen diameter must be increased accordingly.

When a very long screen is needed, which would only be the case in a commercial installation, screen diameter cannot be minimum, or the screen would be too weak. Although a screen should never be loaded with the weight of a string of pipe or casing, there is always the possibility that some loading will occur when the screen is lowered. Also, a very small-diameter long screen is difficult to handle without bending. When a long screen is required, diameter is selected on the basis of minimal stiffness requirements.

Generally the last consideration in well screen selection is yield. Doubling the screen diameter increases well yield by only about 10 percent. This rarely warrants the resulting increase in casing diameter that makes costs go up several times more than double. It also does not warrant the greatly increased costs of drilling the larger diameter hole.

Screen length

Water well yield is directly related to well screen length. Double the screen length and you double the yield in gallons per minute, assuming there is also a sufficient supply. But at the same time, there is no need for yards of well screen to make water available that you will never draw. Screen length is therefore a compromise between your expected future needs, the yield per lineal foot of screen, and the reduction in well efficiency that always occurs with time. Eventually sand will block some slots, encrustation reduce others and corrosion is likely to affect all of them.

Well yield can only be accurately determined by drilling a test well and measuring it. The minimum well screen length can then be determined by extrapolation. This is what commercial drillers do when working in unknown aquifers. If there are any neighboring wells, a yield estimate can be made on the basis of data collected from them. If no information exists or is available and you are drilling blind, some indication of potential yield can usually be gained from formation samples brought up.

Gravel aquifers generally provide the greatest yield. Fine sand is way down on the list and clay approaches zero yield. If you are driving a point or jetting your well and not having any difficulties, you can always withdraw the screen and try another spot if yield is low. But don't be too hasty. Some drillers report that they have been able to increase yield nine-fold in some of their wells by proper development.

For maximum practical yield from a given aquifer of a limited thickness, well screen length must also be limited. The rule of thumb is when the aquifer is less than 25 feet thick, screen length should be held to within 70 percent of aquifer thickness. When the aquifer is thicker than 50 feet, screen length can be 80 percent of aquifer thickness.

It is good practice to position the screen well below the top of the aquifer, for two reasons. First, the deeper the screen within the aquifer the less chance there is for fine sand to seep down into it as water is withdrawn. Second, if the water table drops, a lower screen stands a better chance of remaining within the water supply.

INSTALLING WELL SCREENS

When you drive a well point you drive the screen, too, as part of the point. The same is true of a well jetted with a self-jetting well point. Once these points are down to the desired depth the well screen is in place. When wells are constructed with other techniques, screen installation is accomplished differently.

The well screen in these cases is placed after the hole has been drilled to its final depth, or sometimes a short distance above. During the process, considerable care must be exercised to keep accurate records of pipe length, cable length, and hole depth. You cannot "feel" the bottom of a hole when you have 50 feet or more of pipe and screen on your cable. Any mistake can cost a lot in time and materials.

The simplest, most dependable and most used way to installing a screen is called the pull-back method (FIG. 17-3). The casing is sunk the full depth of the hole and any sand and debris is removed by bailing. A bail plug is attached to the bottom of a telescoping screen and a lead packer fitting is screwed onto the top. The screen is lowered into place with the sand line, which is connected to a hook that engages a loop on the inside of the bail plug.

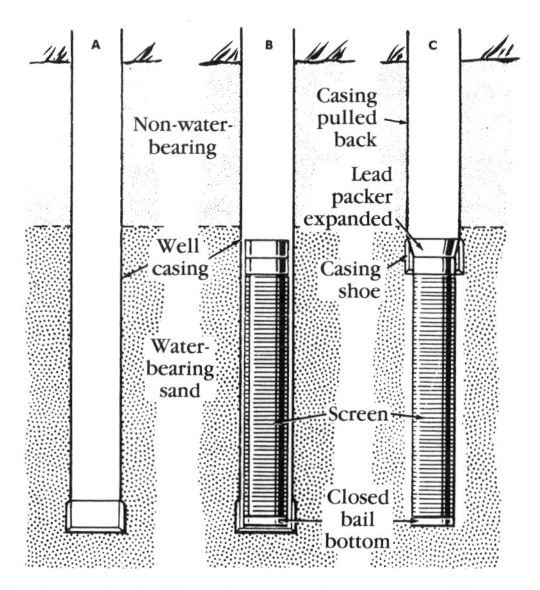

17-3 The pull-back method of installing a well screen. A. The well is drilled and cased to the bottom. B. The well screen is inserted. C. the casing is pulled back. The lead packer is expanded and the screen is permanently exposed to the water-bearing stratum.

Next comes the step that requires you to know exactly how much of the casing is in the ground. The casing is pulled back a foot or more short of the top of the screen; a portion of the screen remains within the casing. This is done by installing a casing ring or a clamp on the exposed upper end of the casing. Hydraulic jacks set under the ring or clamp are used to jack the casing upward.

The sand line is then jiggled to disengage the hook, and is removed. Next, the lead packer is expanded to seal it and the screen inside the casing. This is done with a few light taps of the *swage block* (FIG. 17-4) or, as it is sometimes called, a *swedge*.

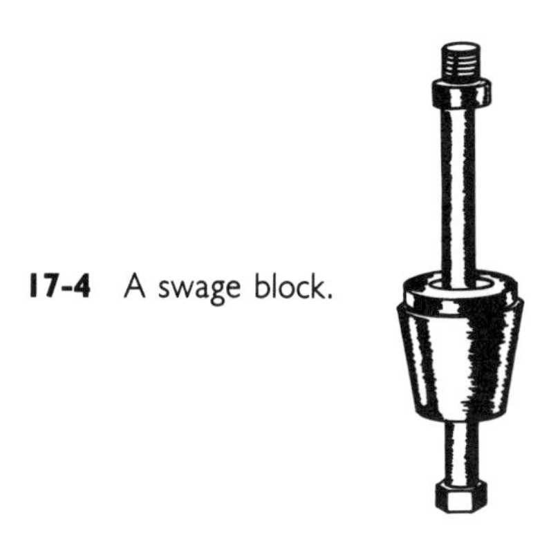

17-4 A swage block.

Bail-down method

This method begins by setting the well casing to a depth that is a foot or more below the desired top of the screen, once that is in position. Special fittings for the screen are required. One type consists of a bail-down shoe and guide pipe that are screwed fast to the bottom of the screen (FIG. 17-5). A pipe is then passed through the screen and connected to a special nipple with one left-hand and one right-hand thread. This nipple is screwed onto the bail-down shoe (FIG. 17-6).

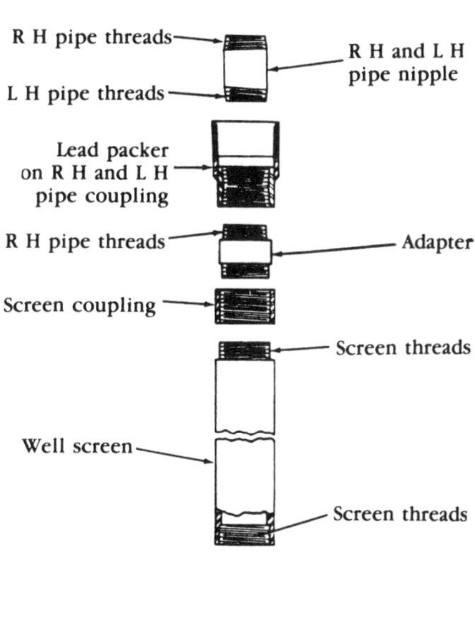

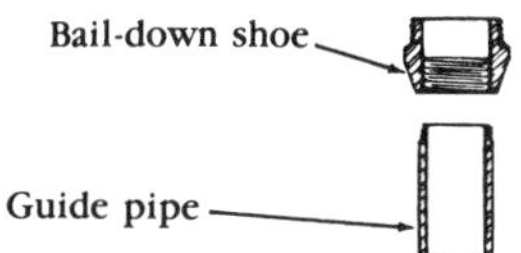

17-5 This screen is fitted with a bail-down shoe and a guide pipe at the bottom. Note the specially threaded fittings at the top.

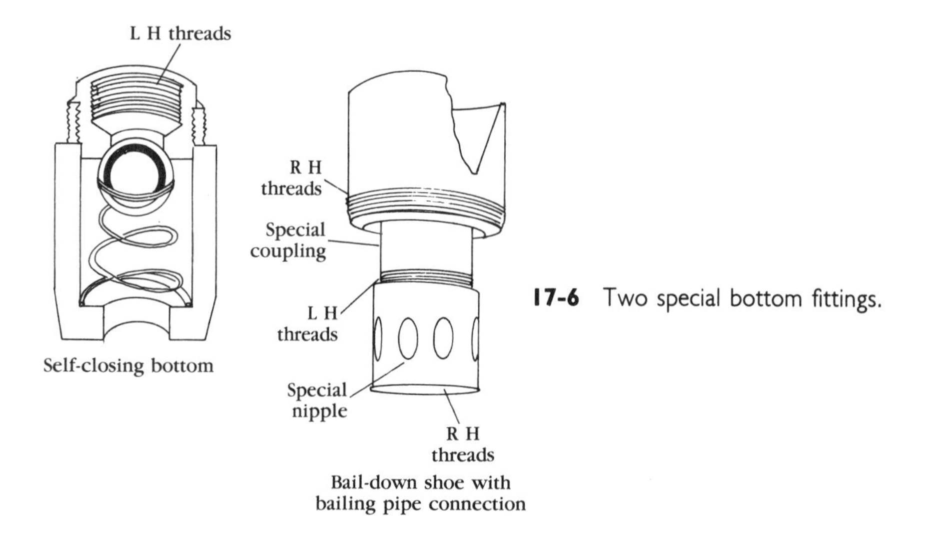

17-6 Two special bottom fittings.

The pipe or pipes are lowered through the casing until the shoe reaches the bottom, where it encounters water-bearing sand. Water is forced down through the pipe and out the shoe, which contains a one-way, self-closing valve. The screen is now "jetted" down through the sand for the desired distance (FIG. 17-7).

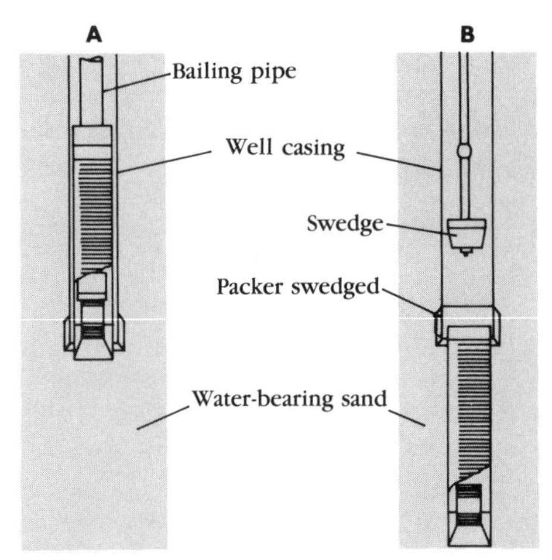

17-7 Steps in the bail-down method of installing a well screen. A. Water pumped out the bottom of the bail-down fitting washes a path for the screen down through the sand. B. The casing has been pulled back and the swedge, or swage, has been used to expand the packer.

The water is turned off. The sand outside the screen, along with that which has been driven up the annulus around the casing, sinks back and locks the screen in place. The pipe attached to the special nipple is rotated counter clockwise, releasing it so it can be removed.

The value of this method is that no more than the final length of casing has to be used, and the last distance at the bottom of the well does not have to be drilled.

Setting pipe-size (ID) screen

When working in non-caving formations, the cost of the well is considerably reduced with pipe-size screen because a casing and inner well pipe are not required. All that is needed is a single string of pipe reaching from the screwed-on screen to the surface. In many instances, however, the bore hole is not noncaving from the ground surface all the way to the bottom of the bore. In such cases, the caving portion of the formation can be temporarily cased and the casing removed when the well is completed.

To position an ID screen and its attached pipe, the screen is screwed fast in place and then the entire assembly is lowered carefully into the hole (either uncased or partially cased). It is supported by the cable until a foot or so above the bottom; unless the well pipe is less than about 50 feet long, the weight of the pipe should never be allowed to rest upon the screen. Any of several sorts of pipe clamps are used to hold the ID screen and pipe in place until the surrounding soil moves in to lock them in position, or until the space around the pipe can be filled with grout.

Rock formations

When the water-bearing formation consists of consolidated rock well screens are usually not required. Water movement is through fissures and cracks in the rocks and no sand or silt is present. Temporary or permanent casing can be used to reach the rock formation. Then the formation is penetrated, and the diameter of the bore hole will then be reduced a calculated amount. The bit and cable is pulled and the permanent well pipe lowered and driven into the reduced-diameter bore hole in the rock. This locks the pipe in place and precludes the entrance of water and soil from the overburden.

RECOVERING AN IN-PLACE SCREEN

One of the advantages of using a telescoping well screen is that it can be removed from the well without disturbing the pipe or casing. This cannot be done with ID screens.

Screens are removed from wells when the slot openings prove to be entirely wrong for the aquifer. They are sometimes removed when a well has to be abandoned and the screen can be used in a new well. Screens also are removed for replacement when they have become badly clogged or corroded, or when encrustation reduces water yield (in-well use of scale-dissolving chemicals is generally not recommended).

The sand-joint method

The sand-joint method of screen removal consists of making a sand joint (FIG. 17-8), attaching it to the end of the pipe string, lowering the joint to the bottom of the screen, and then pulling both joint and screen up out of the bore.

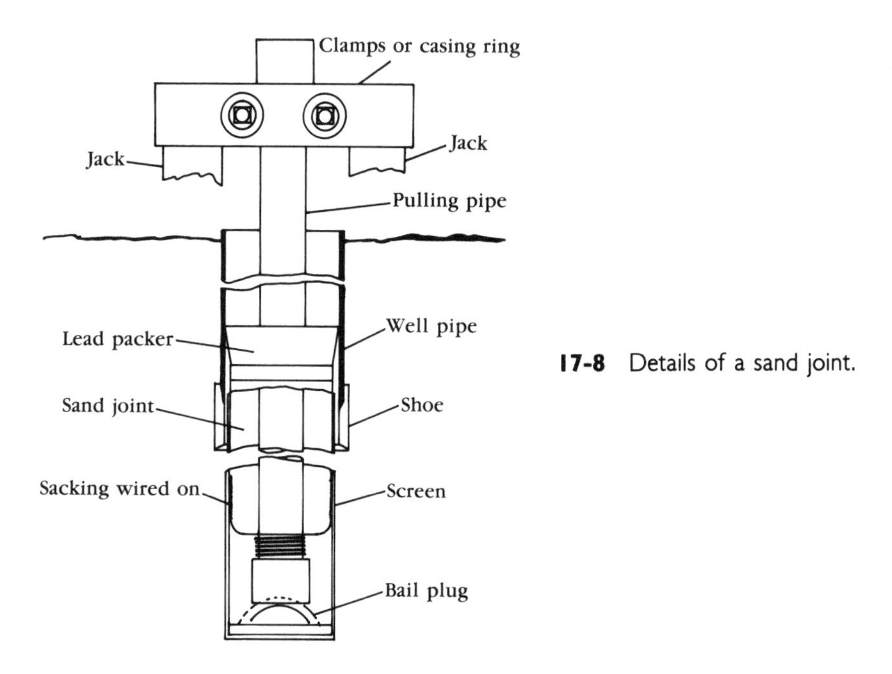

17-8 Details of a sand joint.

As you can imagine, this operation requires considerable force, so not just any pipe can be used in the string. For a screen 3 inches in diameter, use a 1-inch pipe; for a 3½-inch screen, use a 1¼-inch pipe; for a 4-inch screen, a 1½-inch pipe; and for a 5-inch screen, use a 2½-inch pulling pipe.

The force required amounts to many tons. Upward pressure can be provided by a drilling rig or a pair of heavy-duty hydraulic jacks and a strong pipe clamp (or perhaps two, one above the other to prevent slipping).

The sand joint is made by wiring strips of burlap sacking (or an equivalent material) to the pulling pipe just above a screwed-on pipe cap. The spacing is cut into strips 2 to 4 inches wide, depending upon the well screen diameter. The length of the pieces should be such that after one end of a strip has been wired to the pipe, there is a sufficient length left to reach inside the screen and up to its top. The strips are spaced evenly around the pulling pipe.

The sacking, which should form a rough bag with the edges of the strips overlapping, is lowered a short way into the casing. A little sharp, clean sand is poured into the bag, which forces the strips up against the casing wall. The pulling pipe and the bag are then lowered a little, and more sand is added. This process is repeated until the entire bag is inside the casing. The cleanliness of the sand is important, because if there is any clay present the joint will not lock in place.

The joint is slowly lowered to the bottom of the well, where it rests upon the bottom of the well screen. The theory is that as long as the joint is moving downward or is still, the sand within remains loosely held and keeps the bag gently against the casing wall.

When the pulling pipe is slowly and carefully pulled upward, the bottom of the bag tends to expand as the pipe cap presses upward against it. Because sand is almost incompressible, the upward pressure pushes the sand within the bag outward. Thus the sand, which works much better wet than dry, and the bag becomes wedged against the casing. The pipe can then be pulled up, bringing the screen with it.

If the sand joint should have to be removed without displacing the screen, this can be done by washing the sand away. A jet of water or compressed air will do the job.

Acid treatment

Treating the screen of an old well with acid sometimes helps loosen the screen and reduces the lifting force needed to break it loose. The procedure is to fill the screen with a solution of muriatic (hydrochloric) acid and water in a ratio of 1 : 1 by volume. Use a rubber or plastic hose to lead the solution directly into the screen. The acid should be left in place overnight, or at least for several hours. Then the acid and water mix is pumped out before the sand joint is employed.

Chapter 18

Packing and developing

Packing describes placing a layer of gravel just outside the well screen. The layer is usually several inches thick and extends all around the screen and slightly above it. When the packing is properly placed, all water entering the screen must first pass through the packing.

The purpose of packing is to provide a primary, coarse screen in the water flow path. Whether or not it is used is arbitrary, up to the judgment of the driller. Packing is usually not installed in gravel and coarse sand aquifers, but is when the aquifer consists of, or contains a high percentage of, fine sand. The packing stops the fine sand from sifting down to the screen, and enables the use of larger screen slot openings than would otherwise be practical. Larger slot openings increases well yield.

The gravel used for packing is ⅛ inch across and smaller. In the absence of a firm basis for selecting any particular size of gravel, coarse sand is often used.

Gravel sizes larger than coarse sand must be selected with great care. If the gravel is too large for the sand it is meant to retain, the sand will fill the voids and obstruct water passage. That is exactly the opposite of the purpose of installing packing.

Packing gravel must be carefully screened for size so that all the stones are very close in size; no stones larger and none smaller than what is desired must be included. Prepared gravel can be purchased from well-drilling equipment supply houses. To use any size or mix of gravel is worse than useless, because it will consolidate and become relatively impervious to water flow.

Need for packing

As a primary screen to intercept and restrain the movement of fine sand through the well screen, gravel packing works. There is no arguing with that statement. Keep in mind, however, that proper well development is an alternative that can be equivalent to packing. In theory this is possible with all wells, but in practice it is not. Two factors are involved.

One is the nature of the aquifer. Thorough well development will remove the fine sand from around the well screen, leaving coarse sand and gravel, exactly the same condition that artificial packing creates. But in some aquifers this is difficult to do, and in others it is nearly impossible (when the aquifer is fine sand, for example).

The second major factor is cost. Additional material and labor is always required. The quantities of each depend upon the type of well, its diameter, and the method used to place the gravel.

Both considerations must be weighed against the time and effort involved to develop a natural gravel pack. The time required to develop a well naturally can only be estimated on the basis of experience.

So now you encounter the experience factor—the knowledge needed to balance the nature of the aquifer against the costs of artificial development. When a decision has been made to artificially gravel pack a well and the gravel size has been selected, any of the following methods can be used, provided they are suited to the well.

GRAVEL PACKING METHODS

Often called the *positive* method (FIG. 18-1) because it never fails, the *pull-back* method of gravel packing is simple in execution but expensive in materials. The well must be cased, and casing and well-pipe diameters must allow about 2 inches of annular space between them. The packing fits into this space, which dictates the thickness of the packing layer.

While thickness is not crucial, it is important. If too thin, it is ineffective; if too thick, it is nearly impossible to remove the mud present on the walls of the well, a problem in rotary drilled wells that depend upon mud to prevent the bore hole walls from caving. The mud is pressed tight against the wall and forms an almost watertight barrier. When artificial packing is more than a few inches thick, removing this mud is almost impossible.

The pull-back method is simple. The well is drilled full depth and the casing driven to the bottom. The well pipe with the screen attached is set in place at the bottom of the bore and kept centered with a spring guide attached to its bottom end. Then the gravel is gently poured down the annulus. Water can be added to aid the flow of the gravel. Sufficient gravel is added to bring the top of the pack a few feet above the top of the screen. To estimate this distance, the gravel is deposited in measured quantities after the annular clearance has been calculated. The casing is then pulled up enough to expose the screen and the gravel around it.

If you have drilled into a caving formation and there was no alternative, the well casing is already in place and little is wasted beyond the discarded (or reused) upper section of the casing.

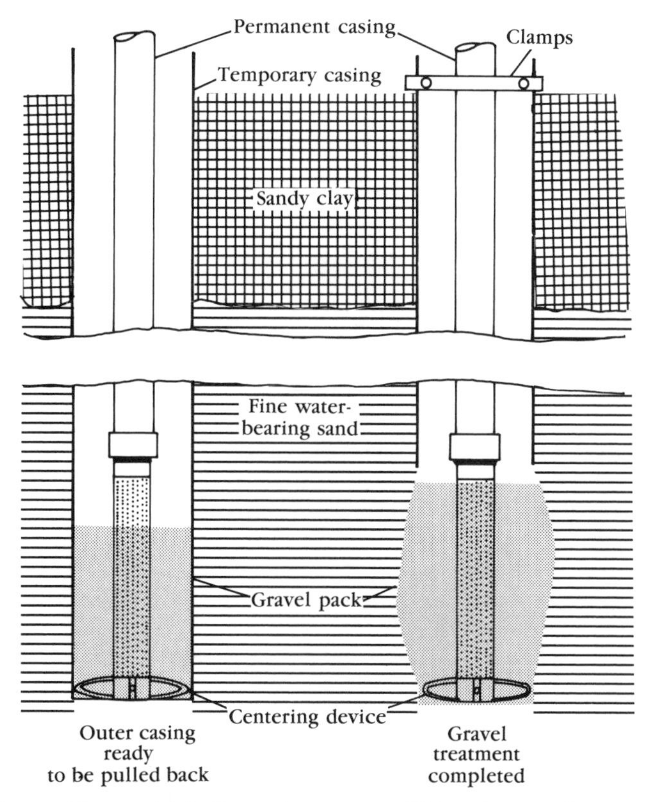

18-1 The positive method of gravel treatment. The casing is pulled back after the gravel has been placed.

Another approach to artificial packing can be employed when the well is sunk by the bail-down method. This involves using a bailing shoe somewhat larger than the well screen (FIG. 18-2). As the screen is bailed down into the sand aquifer, gravel and coarse sand are placed inside the annulus. The packing material follows the screen down.

Noncaving formations

In noncaving formations where the lower reaches of the bore hole do not tend to cave in and can be held in place by relatively clean water alone, the gravel can be positioned by means of a *termie.* This is a long pipe with a funnel at its top end. Its bottom end is positioned in the annulus. Gravel poured into the funnel is guided directly into position. This prevents the gravel from bridging and jamming up somewhere along its journey downward, and the termie also prevents the falling stones from knocking down dirt and rocks during its passage. The goal is to have a single-size blanket of gravel around the screen. Any unwanted addition reduces its effectiveness.

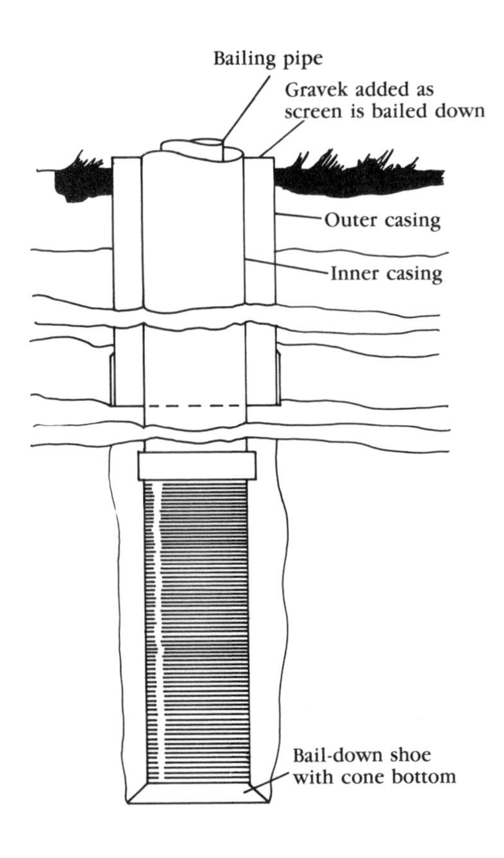

18-2 Packing gravel can be dropped through the annulus when a screen is washed down into place.

Slot size

When the well is artificially packed, it is important to select a slot opening that retains 75 to 90 percent of the gravel. This means that the slot openings will be comparatively large, and much more so than when the goal is to keep out fine sand.

Fine sand alternative

When the aquifer is composed only of fine sand, natural development will not work. However, artificial packing is only one answer to the problem. Another solution is to use a screen with fine slots and at the same time increase the length of the screen proportionally. This might not be practical in a commercial installation where the screen might already run 20 or 30 feet long, but it is practical for a residential well driven or jetted without a casing.

DEVELOPMENT

The purpose of natural development (refer back to FIG. 17-2) is to create a gravel pack immediately outside and around the well screen. Constructing a well by any method always disturbs the aquifer. Some methods cause only moderate disturbance, but at the other extreme, rotary drilling with mud not only disturbs the aquifer, but also seals off all the voids in the sides of the bore hole, considerably reducing water movement into the well. The first goal of development is to return the aquifer to its original condition by removing the barriers to water flow that were produced by drilling.

The second goal is to remove any fine sand and silt present in the immediate area around the well screen. This has to be done, or the well will be a *sand pumper*; the sand and silt will find its way into the water. Otherwise the well screen openings have to be so fine that well yield is reduced unacceptably.

In the normal course of events simply pumping the well will not bring about the desired results. True, in time the worst well will probably run clear. The action of the pump drawing water from the aquifer around the screen will eventually pull all the fine sand and silt into the well.

The trouble with that can be summed up by the term *bridging* (FIG. 18-3). When the grains of sand move in one direction only, they tend to clump up and form "bridges" across the slot openings and across the voids between pieces of gravel. Simply drawing water from the well can clear the water, but it almost always will result in a decreased yield. What is needed is a vigorous in-and-out flow of water; a strong, cyclic flow reversal for as long as it takes to produce the required natural packing.

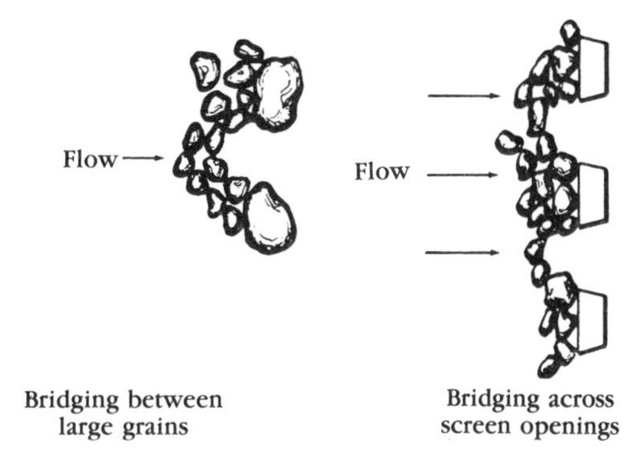

18-3 How grains of sand can bridge openings.

Surge plunger

A surge plunger (FIG. 18-4) is a kind of piston that is moved up and down inside the well pipe. It can be made on the job or purchased. One type has a valve on the bottom that reduces the effectiveness of the piston action, a desirable situation at the beginning of development.

The plunger is weighted to keep the cable taut. It is lowered into the well until just a few feet above the screen. Then it is raised and lowered, slowly at first, for 3 to 4 feet. This is done for several minutes. Then the plunger is removed and whatever sand has been drawn into the

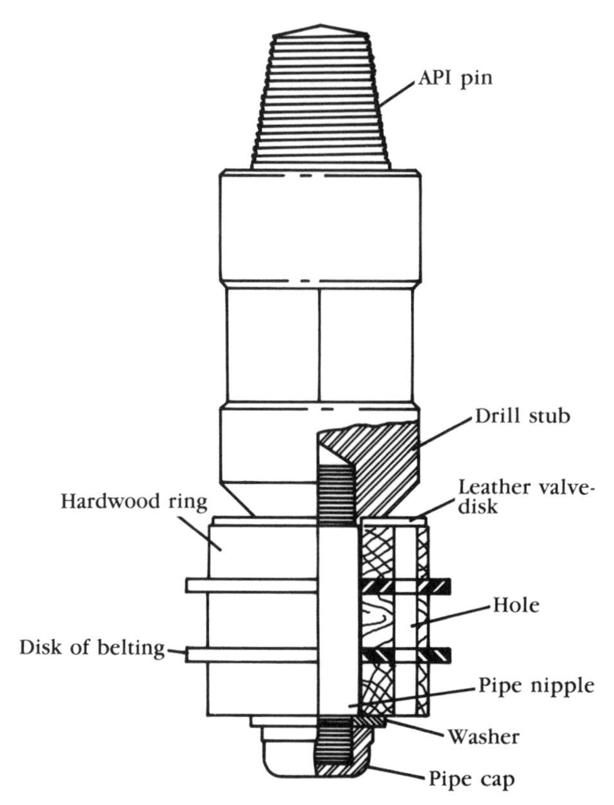

18-4 A commercial surge block. The same action can be duplicated with a simple block-and-leather disk arrangement.

well is removed with a bailer. The process is repeated, a bit faster the second time around.

The quantity of sand removed after each surging operation should be noted to track the progress of development. The surging time should be increased as the amount of sand removed from the well decreases. Surging is stopped only when no more sand can be brought up.

This process can take as short a time as a few hours or as long as three or four days. Everything depends upon the size of the well and the nature of the aquifer.

Developing with compressed air

In this development method water is driven in and out of the well with blasts of compressed air (FIG. 18-5). A pressure of about 100 psi or more is required, along with a volume of anywhere from about 250 to 600 cfm (cubic feet per minute). Depending upon the depth of the well and the quantity of water above the end of the air pipe, 100 to 150 gpm of water can be pumped out of the well by air lift.

In addition, a drop pipe, a compressed air tank and a quick-acting, two-way air valve are required, along with some flexible air line and

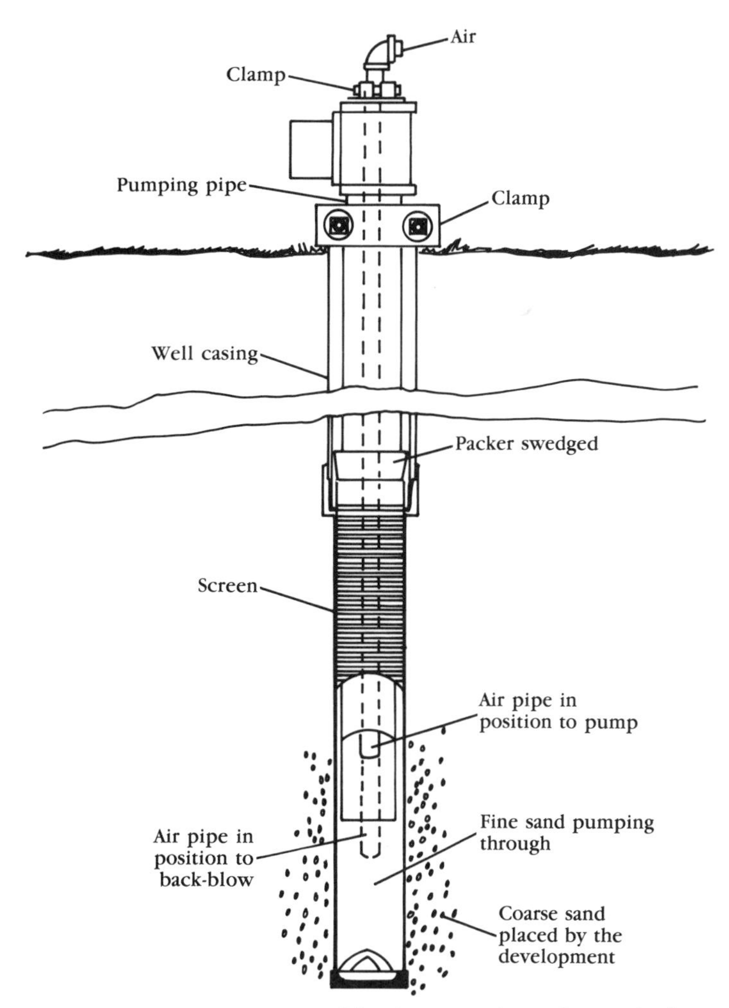

18-5 Developing a well by the open-air surging method.

piping. When a 4-inch or larger well is being developed with air, the drop pipe should be 2½ inches in diameter. The air line should be ¾-inch or larger.

The drop pipe is positioned within the well with its lower end about 2 feet from the bottom of the screen. The air hose or pipe is placed within about 1 foot of the bottom. Sacking is wired around the bottom of the air pipe where it enters the well pipe, to keep water from shooting out the gap.

The first step in the process consists of starting the compressor and turning the valve to fill the air tank to maximum pressure. Then the valve is switched to direct air down the air pipe. The air emerging from the bottom of the air pipe pumps water and sand out of the well. When the water runs fairly clear, the valve is turned back to direct the air in the tank down the pipe. This produces a blast of air that drives the water inside the screen back out and into the formation.

The air tank is then recharged with air, closed off, and the air used to pump the well again. The whole sequence is repeated as often as necessary to move both the drop pipe and the air pipe up a little each time to expose a different portion of the screen to the air blast.

This method will work only when there is a considerable head of water above the end of the air pipe. With only a little water, the air blast will simply push the well water up the casing.

Another method of developing a well with compressed air depends upon pressure being built up inside the well to drive the water out through the screen (FIG. 18-6). This method requires a compressor, drop

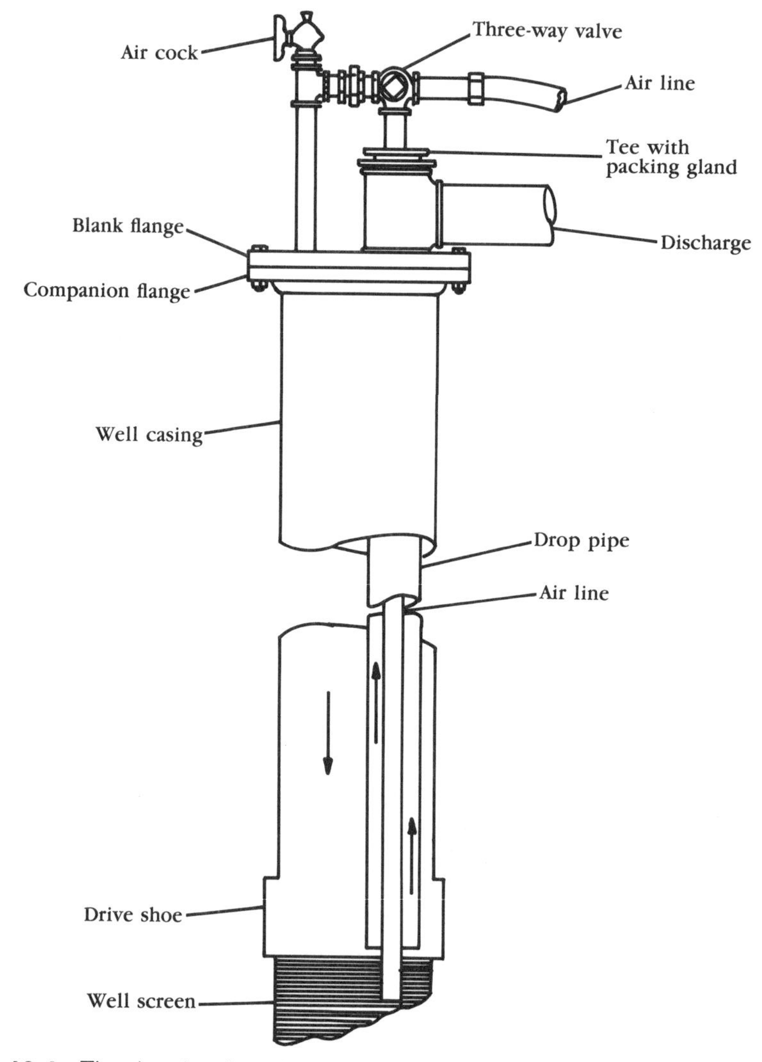

18-6 The closed-well method of developing a well using compressed air.

pipe and air line, a means of temporarily sealing the top of the casing, an air cock and a three-way valve.

The drop pipe and air line are lowered into the well all the way into the screen. Compressed air is introduced into the air line, and this drives the water and accompanying sand up out of the well. The air cock is closed, as is the water discharge pipe. Then the three-way valve is used to direct air down the well casing.

As air pressure builds up within the well casing, the water at the bottom is pushed out through the screen. The air supply is shut off, the air cock is opened, and the water discharge pipe is also opened. The underground water now refills the well. When air stops hissing out of the air cock, the well is as full of water as it will ever be. The process is then repeated again and again until air pumping brings up nothing but clear water.

Development by backwashing

Because any cyclic motion of water in and out of the well screen will produce a natural gravel pack around the screen, any arrangement that does this will produce the desired results.

One technique consists of rapidly pumping the well dry, or almost so, and then letting the pumped water flow back down in a rush. Another method involves using a jet of water to produce the desired backwash (FIG. 18-7). Still another technique consists of building water pressure within the well and then suddenly releasing it.

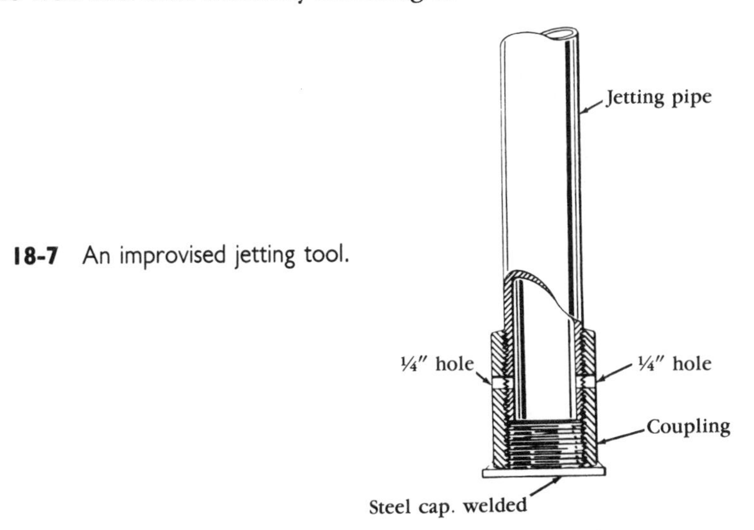

18-7 An improvised jetting tool.

Developing rock formations

All the previous methods of developing a well can be used in rock formations. It is customary to increase the size and number of cracks and fissures and other openings through which water enters. An acid that dissolves limestone is used.

Redeveloping old wells

With the passage of time, well yield drops due to encrustation and scale deposits if that is the nature of the water. If the water is acidic, the well screen will be eaten away. Encrustation and scale deposits are self-growth phenomena. When the slots first decrease in size, the water flow slows. As the flow slows, more minerals are deposited, further reducing the slot area and the water flow, and so on. Eventually the slots can become completely closed.

Any of the aforementioned development methods can be used to increase the yield of an old well. If done early enough, scale and encrustation can often be loosened and washed away. If these methods do not work, an acid or chemical treatment may be used provided that is not specifically recommended against.

Sufficient muriatic acid is released by means of a pipe to fill the space inside the screen. The acid is permitted to remain in place for an hour or two and then the screen is lightly surged. Some two to four more hours are permitted to pass, and the well is bailed or pumped free of the acid and surged some more. Finally the well is bailed free of whatever debris the acid and surging has brought into the screen.

If the acid treatment does not produce results, try a solution of hexametaphosphate. Mix a sufficient amount of the chemical in a 50-gallon barrel to produce a solution of 3000 to 5000 ppm (parts per million). Or, use 4 pounds of the chemical per 100 gallons of water standing in the well. Also add calcium hypochlorite to produce a solution of 50 to 100 ppm of the chlorine. Permit both solutions to remain in the well for an hour, then surge it vigorously for 3 hours and pump the solution and debris out. Repeat this procedure three or four times.

If none of these procedures increases the yield satisfactorily, there is no alternative but to remove the well screen and either clean it or replace it with a new one. If an old well has started to pump sand, then the screen has corroded through and must be replaced.

Chapter **19**

Completing the well

A prime concern when selecting the site of a well is the possibility of pollution. That is why all possible sources of contamination must be carefully considered before well construction is undertaken. But proper siting is not enough by itself. The well must also be protected from surface water contamination that might enter by draining down through the annular space around the well pipe or casing, finally reaching the well screen. Another concern is surface water that leaches through the earth and moves sideways to reach the casing and travel downward.

GROUTING

Grouting is a time-tested way to seal the annular space around a well pipe or well casing with a cement slurry put in placed under pressure. This not only seals the space between the bore and the well pipe or casing, it also seals off cracks and crevices in the formation and helps protect the pipes from corrosion. Some commercial well drillers have reported that properly pressure-grouted casings have withstood corrosion for 30 or more years. Pressure grouting is also used to seal off undesirable aquifers in situations when it is necessary to drill through one aquifer to reach another and the upper aquifer is contaminated or filled with unpotable water.

Pressure grouting, when properly performed, limits the water that enters the well screen to the water present in the aquifer in which the screen is immersed. No other water can enter. To ensure this, the grout must extend several feet below the lowest depth to which the well will be pumped (FIG. 19-1).

The technique or method you select will depend upon the depth of the well and the nature and size of the bore hole. No particular method is best for all wells, and the driller has to improvise and adjust to the individual conditions.

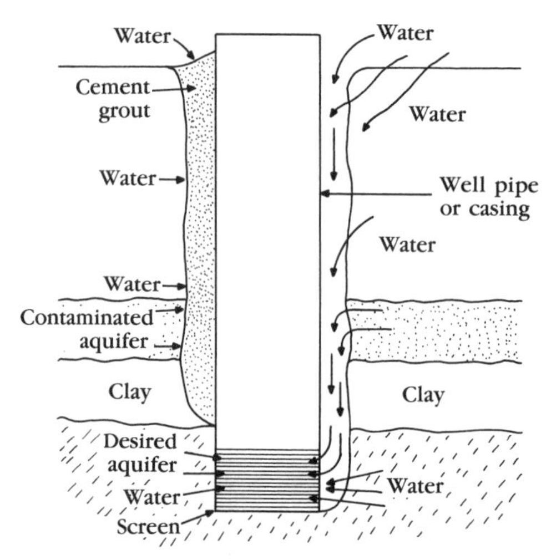

19-1 The left side of the well shown here has been properly sealed with cement grout. Neither surface water nor water from the contaminated aquifer can get into the well. The right side has not been properly sealed. Surface water and contaminated water can enter the well.

A driven well cannot be grouted because there is no open space between the riser or well pipe and the surrounding formation. The upper few feet of pipe can be exposed by hand digging and the open space filled with a mortar mix. You can use a mix of 2 parts sand to 1 part portland cement. The same arrangement can be used with a jetted well.

When the bore hole is self-supporting from the surface to the top of the well screen, the screen must be protected from being blocked and sealed off by the grout. This can be done by filling the annulus around the screen and up to several feet above it with graded gravel of the proper size relationship to the screen slots. In effect, this is artificial packing. If the screen has already been packed, the packing should be topped with another foot or so of gravel. This can be gravel of any size or mix. Other sealing means are also available.

If the screen has been exposed by the drawback method but has not been packed, give the aquifer time to consolidate around the screen. To speed the process, the well can be pumped. To be certain no grout works its way into the screen, several vertical feet of gravel can be placed within the casing atop the lead packer. If the screen is deep within the aquifer these precautions are unnecessary because grout, even if thin, will not seep through more than about 1 foot of small gravel, and even less through coarse sand.

To grout a well use a mixture of portland cement and water on the order of 1 bag of cement to 5½ to 6 gallons of water. To obtain a more free-flowing mix, add about 2 to 6 percent of finely ground bentonite. This special clay, often added to drilling mud, will also reduce the shrinkage normal in portland cement mixes.

The grout must be thoroughly mixed and run through ¼-inch mesh screening (hardware cloth) to remove lumps. Mixing can be done by hand or with a power concrete mixer. You can estimate the required quantity by calculating the volume of the annular space and adding a percentage for waste and error—10 percent is a common figure.

Above ground on a warm day, a portland cement water mixture will set in about 1 hour, harden in about 36 hours, and cure to 75 percent of its ultimate strength in about 1 month. In the ground, where the temperature might be in the 50s and 60s, the initial setting could take several hours.

Position the grout by using a termie pipe no smaller than ¾-inch diameter and preferably larger (FIG. 19-2). Place one end of the termie just above the top of the packing. Connect the other to a pump capable of driving the grout down the termie at a pressure of 100 psi or whatever is considerably greater than the head (pressure) of the water at the bottom of the well.

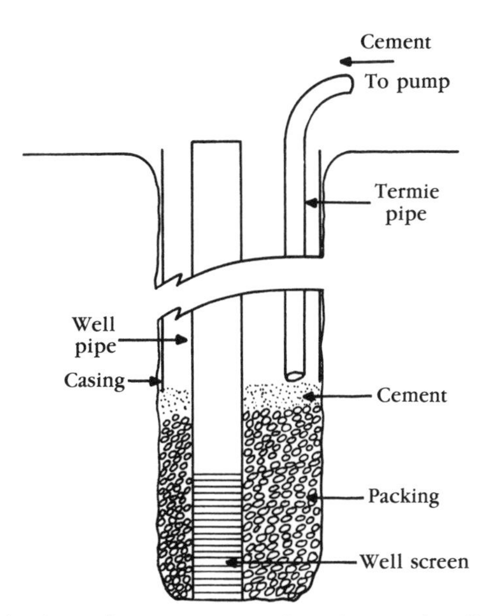

19-2 A termie pipe is used to pump grout into the annulus. Packing around the screen keeps the liquid cement from sealing the screen.

Just pouring the grout down the well will result in sand. The cement has to be deposited under pressure so the water is moved up and displaced by the grout. Because the grout should extend below the water level in the well when pumped at maximum capacity, the end of the termie will always be under water. The termie is raised as the annulus is filled (FIG. 19-3). Pumping is stopped only when the grout begins to run out the annulus.

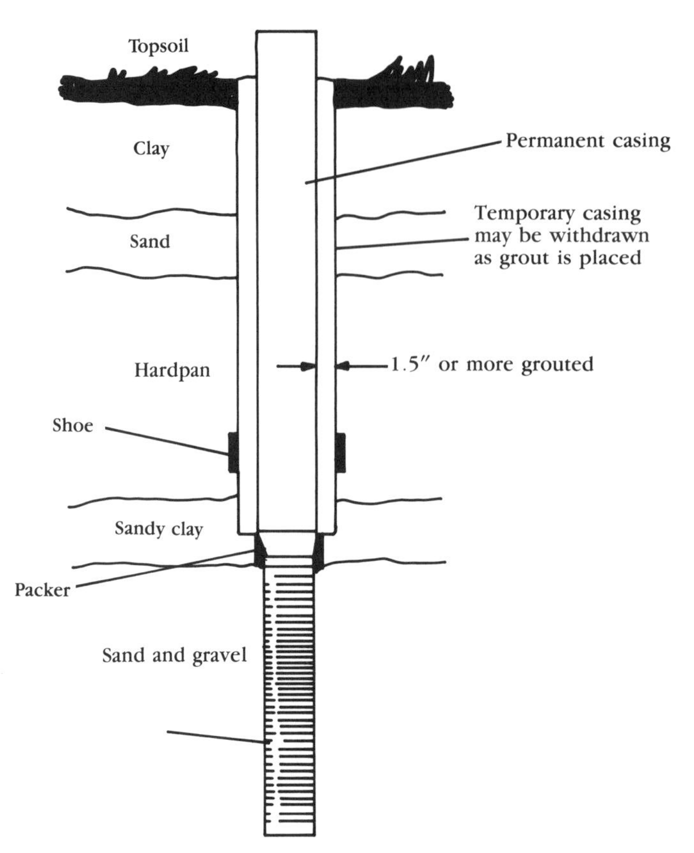

19-3 The bore hole is drilled large enough for the inner casing (well pipe) and an outer casing. The screen is positioned and the packer is expanded. The space between the well pipe and the outer casing is grouted. As this is done, the outer casing is slowly removed.

DISINFECTION

In the course of constructing any well, the pipes and the tools you use will always pick up some soil, which invariably contains some pathogens. This means that any well, after being constructed or after any kind of work has been done, must be disinfected before it is fit for human use.

The simplest method is to use a chlorine solution. Generally sodium hypochlorite is used. This is prepared by bubbling chlorine gas through a solution of caustic soda. As a practical matter, common household bleach containing 5 percent available chlorine makes a good choice. To sterilize a well, a solution containing about 100 ppm of available chlorine should be used. To obtain this concentration, mix 0.4 quart of 5 percent bleach to every 100 gallons of water in the well and associated piping and pump.

The chlorine must be agitated in the well and pumping equipment and should be permitted to remain in place for four hours or more. It is also necessary to disinfect all gravel pack material. It is good practice to pour a little of the chlorine into the well while the drilling tools are in place. If the well is very deep, it will be difficult to achieve the required chlorine concentration with a liquid. In such cases dry calcium hypochlorite is placed in a suitable container and lowered to the bottom of the well. The container is moved up and down until the chemical dissolves.

To make sure the entire water system is being disinfected, the chlorinated well water should be pumped through the plumbing system and all the faucets opened until the smell of chlorine is obvious. Then the faucets are closed and the chlorine permitted to do its work. After a time, the water is pumped through the system until the smell disappears and the water no longer feels soapy. The system is disinfected and ready for regular use.

Chapter 20

Water treatment

Once a well has been built, tested, disinfected and put into service, often nothing further is done to treat the water if the original analysis was satisfactory and the water is safe to drink. But raw water is not always satisfactory—it can be unsatisfactory without being unsafe—and water that is initially suitable for consumption does not always remain so. Surface waters are consistent in their variability. Ground water is less likely to harbor pathogens, but more apt to contain undesirable tastes or odors or a variety of mineral impurities.

The object of treatment is to raise the quality of the water to the highest possible level. The procedures incorporate, modify or supplement various natural processes. The first and most important goal is to free the water from any pathogenic organisms or other undesirable materials, like heavy metals, that might constitute a health hazard for the users. The second goal is to condition water, reducing levels of or eliminating completely, any chemicals or other impurities that are anesthetically objectionable.

SEDIMENTATION

Sedimentation is a process whereby the water is stilled for a period of time to allow heavier particles that are temporarily suspended to settle out by gravity.

A pond can be used for this purpose, but in an individual water system a closed tank or basin of proper design is best. The water must be still for at least 24 hours to accomplish any worthwhile reduction in suspended matter. This means that the tank must contain several times the expected daily consumption of water. Inflow must be arranged so that the water flows and spreads out more or less uniformly across the whole tank surface, toward the outlet. But the flow must also be baffled so that the content of the tank does not become agitated and roiled up again by sudden inflow or outflow.

To facilitate cleaning, two tanks or a double compartment arrangement can be constructed. The two can be used independently and emptied and cleaned alternately, so there is no time when water is unavailable.

COAGULATION-FLOCCULATION

Coagulation means to gather something together into a mass. A floccule is a tiny bit of material suspended in or precipitated from a liquid; flocculation is a process of aggregating floccules into a mass. In water treatment, this is accomplished by means of coagulation techniques.

For example, alum, which is hydrated aluminum sulfate, can be added to turbid (full of roiled sediment) water. The water is allowed to stand, which precipitates the floccules and they combine into a flocculent mass sometimes called a floc. The mass settles out and can be removed as a sediment during periodic tank cleaning.

This process, although effective, would be used only in unusual circumstances in an individual water system. It should be done only with competent engineering advice.

FILTRATION

Filtration is one of the most common methods used to treat water in individual systems. There are many misconceptions about the efficacy of filtration, and unfortunately many homeowners are talked into installing, and paying dearly for, systems that are marginal, of no help for a particular condition, or just completely useless. Even worse, some small household filters can actually introduce bacteria into otherwise safe water by way of outside contamination. And others that are sometimes sold as water "purifying" filters simply will not purify water in the sense of producing water that is bacteriologically safe and free of pathogens. Competent professional advice, and perhaps consumer advice as well, is needed in this area.

The purpose of filtration is to remove suspended material from the water as it passes through beds of porous material. Particulates will be removed; bacteria will not. If the water is very turbid, especially if nearly opaque, it should be partly cleared by sedimentation before it is filtered. There are several kinds of filters in common use.

The *slow sand* type consist of a bed of special fine filter sand 4 feet deep or more and resting upon a 1-foot bed of gravel. Water passes through the sands at an optimum rate of 0.05 gallons per minute per square foot of filter area and a quantity of 60 to 180 gallons every 24 hours.

Pressure sand filters are similar to slow sand filters but are pressurized to force about 2 gallons per minute, sometimes more, through the sand beds. Provisions for periodic manual or automatic backwashing, which flushes the sand beds clean and prevents them from clogging up, are a part of this system. This kind of filtration is not suitable for small individual water systems because of the high cost of equipment and the need for a fair amount of attention from an operator. It is an effective means for several households installing a small community or subdivision water system.

Diatomaceous earth filters contain a quantity of that material. Water is forced through the filter under pressure in similar fashion to the pressure sand filter system. Filtering capability and water flow rates are also about the same.

Porcelain, ceramic or equivalent filters are widely sold for household filtration. Most are point-of-use type, made to attach directly to the faucet. They are ineffective, can foster bacterial growth and should be avoided.

Pad filters appear in a variety of shapes and sizes. Each unit consists of a canister of some sort containing one or a series of pads, spools, wads, or sheets of fibrous filtering material. They are low-volume, low-capacity devices good only for filtering out large particles. The elements must be frequently cleaned or replaced, and operation can be expensive without affording particularly good filtration.

DISINFECTION

Disinfection of the well after completion but before putting it into service was discussed in Chapter 19. In many instances this is the only disinfection ever done, but the entire water system should be disinfected every time it is opened up for any reason, as for service or repair. In addition, more and more locales are requiring that well water, even in small, individual systems, be disinfected on a continuous, automatic basis as a precaution against health hazards.

This is the most important of all water treatments. The purpose is to destroy all pathogenic bacteria and other harmful organisms. The turbidity of the water being treated must be low—a minimum of suspended particles should be present. After disinfection, the water must be kept in a closed system to prevent any recontamination from outside sources. There are several ways in which water can be disinfected but because compounds of chlorine work so well and are so readily available everywhere at low cost, chlorine is by far the most widely used disinfectant.

To discuss chlorine disinfection, you need to be familiar with several terms:

- *Chlorine concentration* is the amount of chlorine held by a certain amount of water. This is expressed in either milligrams (of chlorine) per liter (of water), mg/l, or as parts per million, ppm. The two are interchangeable for this purpose
- *Chlorine dosage* is the actual amount of chlorine introduced into the water.
- *Chlorine demand* is the amount of chlorine taken up by substances in the water. Some of the chlorine introduced into the water combines with impurities like ferrous iron or nitrites and becomes "locked up" and unavailable for disinfecting purposes.

- *Free chlorine* is the amount of chlorine left in the water after the chlorine demand has been met. This is the most useful and effective chlorine for disinfection and is called the *chlorine residual.*
- *Combined chlorine* results when there is ammonia nitrogen in the water supply. Some of the chlorine that is introduced combines with that substance to form other chlorine compounds that have a modest disinfecting capability.
- *Chlorine contact time* is the period that elapses between the initial introduction of the chlorine and the time the water is used.

Usable chlorine

There are numerous compounds of chlorine, but the ones most commonly available and widely used are sodium hypochlorite and calcium hypochlorite. Both work very well to disinfect small water supplies. The use of household bleach for well disinfection has already been discussed; it can also be used for emergency disinfection of the water system, as after a repair job. For continuous treatment, however, usually tablets or powder are purchased in bulk and made up into solution as needed.

Calcium hypochlorite can be had in either form and is available from chemical and water system supply houses in cans or drums. The material is stable and stores easily. The same is true of the sodium hypochlorite available at chemical and swimming pool supply outlets in solution form, provided it is stored in a cool, dark spot. Tablet and powder forms contain from 65 to 75 percent available chlorine by weight, while the solutions range from 3 to 15 percent available chlorine by weight; household bleach typically runs 5 percent.

Powder and tablet hypochlorite must be prepared in solution at frequent intervals because it degrades fairly quickly. Prepared solutions are often used full strength, powder and tablet forms are reduced considerably. The strength of the solution to be used is determined by the kind of equipment employed to introduce it into the water and the rate of the water flow.

The chlorine content of the powder or stock solution has to be considered, according to weight when making up the solution. To find the weight of compound needed for a particular quantity of solution, multiply the percent strength of the required solution times the number of gallons required times 8.3 (weight of water per gallon) and divide that by the percentage of available chlorine in the compound. Thus, if you need 5 gallons of 5-percent solution and your chlorine compound supply contains 70 percent available chlorine:

$$5 \times 5 \times 8.3 = 207.5$$
$$207.5 - 70 = 2.96 \text{ pounds}$$

Dissolve 3 pounds of compound in 5 gallons of potable water and you will have just about the right mix for disinfection purposes.

Chlorination equipment

There are several kinds of equipment that will automatically introduce chlorine solution into a water supply, collectively called *hypochlorinators.*

One of the most common types is the *positive displacement feeder.* It is reliable and feeds accurately, is readily adjustable, and is composed mostly of a simple diaphragm or piston pump. It can be set up to operate in concert with the water pump.

An *aspirator feeder* always operates when the water pump is running, for it is the vacuum created by water rushing through a venturi that sucks a metered amount of chlorine solution into the water stream. Solution flow can be easily regulated with a control valve, but pressure variation in the water flow can vary the feed rate. Nonetheless, this is a simple, practical and reliable system. There are several design variations.

Some *suction feeders* operate on the siphon principle and the chlorine solution is introduced directly into the well through a tube whenever the water pump operates. Other designs use a chlorine supply tube running directly to the suction side of the water pump. When the pump operates, the chlorine is pulled from the supply tube as the water is drawn from the well.

A *tablet hypochlorinator* is less reliable than those just discussed and requires more attention and maintenance, but has the advantage of not requiring electricity for its operation and it operates effectively in low water pressure situations. A closely controlled injection of water into a tablet bed slowly dissolves the tablets and provides a continuous supply of fresh chlorine solution.

Gas feed chlorinators operate by injecting measured quantities of chlorine gas from a pressure cylinder into the water supply as the water pump operates. This is an expensive system that must be used with stringent safety precautions, and is neither feasible nor recommended for individual water systems. It can be a workable alternative, however, in a small community or subdivision water system serving several houses.

Testing

To find out if a water supply contains sufficient chlorine, it must be tested. There should be a certain quantity of total available chlorine residual, which is the free plus the combined residual, in the water at any given time. Enough chlorine is being introduced when that amount produces a desired residual after a defined contact period.

The test is a simple one, called a DPD (N, N-diethyl-p-phenylenediamine) colorimetric test. It is done by placing a pill of chemicals in a special test tube full of water. Free chlorine residual produces a violet color that is then compared with a color chart to show what quantity is present. These kits are readily available and inexpensive. Your local health or water department can tell you where to obtain one.

Effective disinfection

The effectiveness of chlorine disinfection depends upon several factors, all of which should be borne in mind as the disinfection system is developed and operated:

- The higher the chlorine concentration in the water, the more effective the disinfection and the faster it will occur.
- The lower the pH of the water, the more effective the treatment will be. The pH is a measure of the hydrogen ion in the water, as well as the acid and alkali content. Values range from 0 to 17, with 7 being neutral; below 7, acidity increases, above 7, alkalinity increases. Test kits are readily available.
- Free chlorine is much more effective than combined chlorine. Thus, a greater amount of free chlorine residual is desirable.
- The longer the contact time of the chlorine with the water, or more particularly, with any organisms in the water, the more effective the disinfection.
- The higher the temperature of the water being treated, the more effective will be the disinfection.

CONDITIONING

There are many causes of perfectly safe water being nonetheless unpalatable or otherwise unsatisfactory for ordinary consumption. *Conditioning* is the process of improving water quality to an acceptable level.

Softening

Softening is the most familiar method of conditioning household water supplies. The process is used primarily to remove or reduce minerals which cause hardness, mainly calcium and magnesium. It may be desirable when: producing a soap lather is difficult; scale forms on cooking utensils; whitish, rock-like formations build up on faucet nozzles, pipe interiors and water tank walls; heat transfer from heating elements in water heaters is impaired; and automatic valves in dishwashers and clothes washers malfunction from scale build up.

The ion-exchange process is used in domestic water softening installations. In most systems, the exchange medium is either a synthetic resin or a gel zeolite which exchanges sodium ions for the calcium or magnesium ions in the water. The medium is regenerated periodically by passing a brine or salt solution through the bed. Only relatively small quantities of water have to be processed because the water produced is at zero hardness. It is then used to dilute incoming hard water to produce a final result of about 3 to 5 grains of hardness per gallon.

A water softening system should be designed and installed only by qualified personnel from a reputable company, using recognized and approved brand-name products. Nearby reliable and responsible service must be readily available, and all materials and workmanship should be guaranteed. A poor installation or bad materials or equipment can cause no end of trouble.

If any consumers of the water are on a restricted sodium diet for medical reasons, they should probably not use water from an artificially softened source. Consult a doctor first.

Iron and manganese

The presence of iron or manganese, or both, is common. The condition is easy to spot: the iron combines with oxygen from the air and forms a reddish-brown precipitate on the walls of sinks, toilets, and tubs. Manganese acts about the same, but the stains are brownish-black. The water can have a metallic taste, as can foods prepared with the water. Stains in fabrics cannot be removed, and bleaching may make matters worse. Mixed drinks can be unpalatable.

The methods used to condition water containing these impurities depend on the specifics of the problem. One process combines automatic continuous chlorination coupled with fine filtration. The chlorine oxidizes the iron or manganese, precipitates it, and the filters strain out the precipitates. Other methods include treating with potassium permanganate followed by filtering; aeration followed by filtration; and ion exchange with greensand. Combinations of methods may also be used. Qualified people are needed to analyze each situation and devise a workable conditioning system.

Iron bacteria

Iron bacteria can complicate removal of iron compounds from a water supply, a fairly common problem. These creatures thrive upon the oxidation of iron to its insoluble form. The result is a slimy, slippery, reddish-brown coating that accumulates on the inside of toilet tanks and similar locations. Over time the slime can build up considerably and cause difficulty. Conditioning is much the same as for iron and manganese compounds, only more so, and more filtration and filter backwashing is required.

Corrosion control

One of the most important aspects of water conditioning in many locales is corrosion control. There are several water characteristics that foster corrosion of pipes, fittings, and water-using or -carrying equipment: high acidity; high conductivity from dissolved mineral salts; a large amount of dissolved oxygen; a large amount of carbon dioxide; and high water temperature.

Corrosion can be a serious problem, leading to the failure and subsequent need to replace the metallic components of the water system. It is an electrochemical reaction whereby the metal deteriorates—is literally eaten away—when in contact with soil, air or water. Correcting the problem requires mitigating the conditions that encourage corrosion through instigation of any of several control measures, such as oxygen removal, installation of dielectric unions, and injection of protective film-forming materials such as silicates. Extensive use of plastic pipes and fittings is another possibility.

Fluoridation

Some individual water supplies are naturally fluoridated, most are not. Because fluoride in the water has been found to reduce dental caries, well-owners may wish to condition their own water. Equipment for home water supplies, even very low-volume systems, is readily available and is reliable and economical, especially when combined with other conditioning services. Follow local or state health department recommendations and maintain the system and test the water regularly.

Aeration

Aeration is a process of bubbling, jetting, streaming, cascading or spraying water through the air by any of many methods. This can be done in the open, but because of possible contamination a completely enclosed and sealed system is strongly recommended.

Aeration can remove odors from water, and will introduce oxygen into otherwise stale or "flat-tasting" water that has stood in storage, enhancing its taste. Strong aeration will eliminate carbon dioxide and other gasses from water and will help oxidize dissolved iron and manganese so they will precipitate and can be removed. Aeration systems should be developed and installed by qualified personnel.

Tastes and odors

Water often has taste or smell that some consumers find objectionable and others do not. Some, such as the taste of chlorine, may be necessary. Most, however, can be combatted by various means. Dissolved gasses are frequent culprits; hydrogen sulfide with its distinctive "rotten egg" odor is one, and is usually called "sulfur water."

The solutions to these problems depend upon specifics. Aeration often will do the job, as will a combination oxidation-filtration process. Chlorination is an effective means of overcoming many odor and taste problems. Activated granular carbon filters are often used because the carbon adsorbs large quantities of dissolved gasses, very fine solids, and soluble organic substances. If the problem is peculiar taste and odor attributable to algae, a common situation in small, low-volume individual water systems, treatment by the addition of copper sulfate to the water at the source is usually effective. Here again, professional advice and help is recommended to work out appropriate conditioning plans.

Chapter 21

Pumping, distribution, and storage

The well is the biggest part of an individual domestic water system, but it is not the only part. You will need some means of drawing the water, transporting it to the house and storing an amount against peak demands. The fundamentals of these subjects are discussed on the following pages, but details have to be sorted out on the basis of each installation. They vary widely, depending upon such localized factors as availability of particular products and services, costs, well specifications, available space, water quantity and pressure required, piping distances, and pertinent regulations.

PUMPING

In a small percentage of individual water systems, the water is delivered from the source by gravity feed, by hydraulic ram or even by lowering a bucket. Usually water is forced into a storage vessel by means of a powered pump. The power source is an electric motor, but if electricity is not available the pump can be coupled to a small gasoline or diesel engine. Even wind power can be effectively harnessed.

Apart from rams and hand-operated pumps, there are three kinds of pumps commonly installed in individual water systems.

Positive displacement pumps operate by displacing the water through a pumping mechanism. One common variety is the *reciprocating* or *piston* pump, which contains a plunger that is driven back and forth in a cylinder by a crankshaft and connecting rod.

Another type of positive displacement pump is the *helical* or *spiral rotor* pump. This consists of a spiral shaft rotating in a rubber sleeve. Water is trapped between the shaft and the sleeve and is forced outward into the distribution line.

A *regenerative turbine* positive displacement pump makes use of an impeller in a raceway. This system forces water through the pump under much more pressure than can be generated by centrifugal force alone.

Centrifugal pumps are the second type of pump in common use. This pump contains an impeller mounted on a shaft turned by the motor. This rotor assembly increases the water velocity and forces it into a casing shaped to slow the flow and increase the pressure. Each impeller-casing set is called a *stage*. Stages can be coupled together to provide added pressure and increase the lift capability. Such pumps are called *multistage* pumps, and there are two kinds in particular that are installed in water systems.

A *turbine* centrifugal pump (FIG. 21-1) consists of a pump unit placed in the well below the maximum water draw down level, a drive motor mounted on a base at the top of the well and a vertical shaft connecting the two. The pump bearings may be lubricated by oil or water, but the latter is preferable to avoid possible contamination.

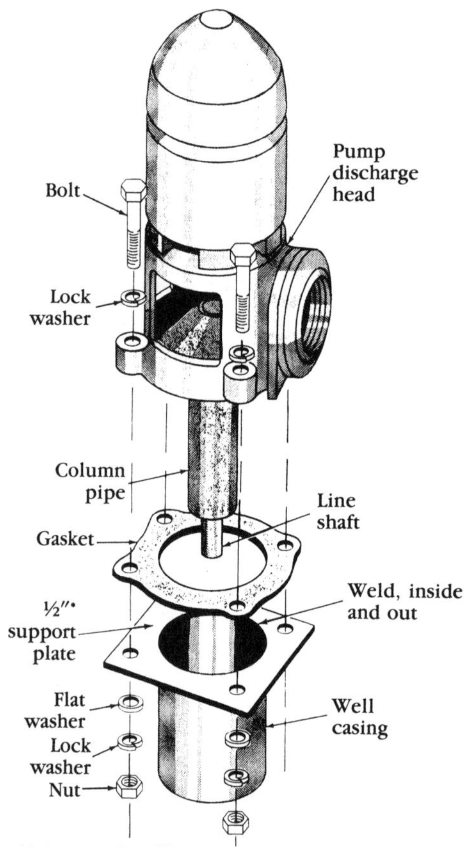

21-1 Vertical turbine pump mounted on well casing. The motor unit is at the top and the pump is at the bottom end of the line shaft, submerged in the well.

A *submersible* centrifugal pump (FIG. 21-2) consists of an electric motor, inlet body, several diffusers and impellers, and a check valve all assembled in one long, thin cylinder. The entire assembly is supported on the discharge pipe, with the aid of a safety line, below the maximum water draw down level in the well casing. The electric cables and the motor must be waterproof and the system carefully grounded and protected against lighting strikes.

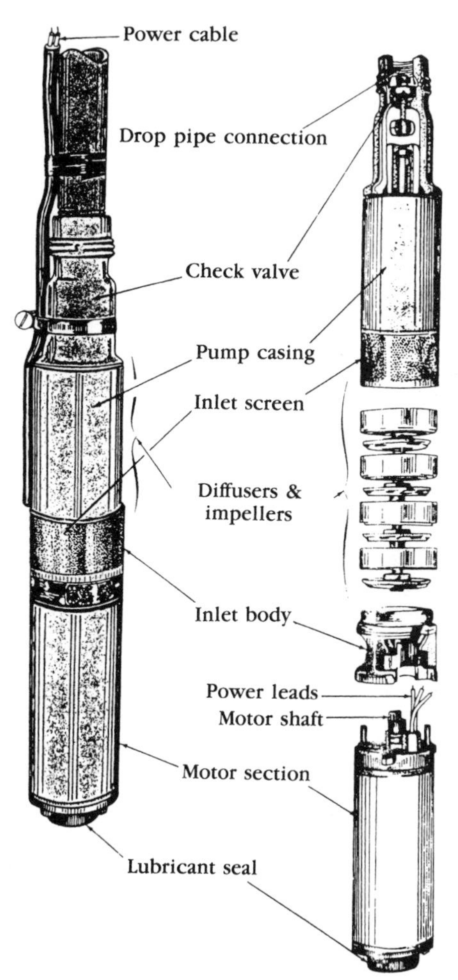

21-2 A typical submersible pump and its component parts.

The third kind of pump in common use is the *jet* or *ejector* type (FIG. 21-3). This is fairly complex system; some of the water from the centrifugal pump is diverted through a venturi, creating a low-pressure zone. Well water enters this zone and the velocity of the water coming

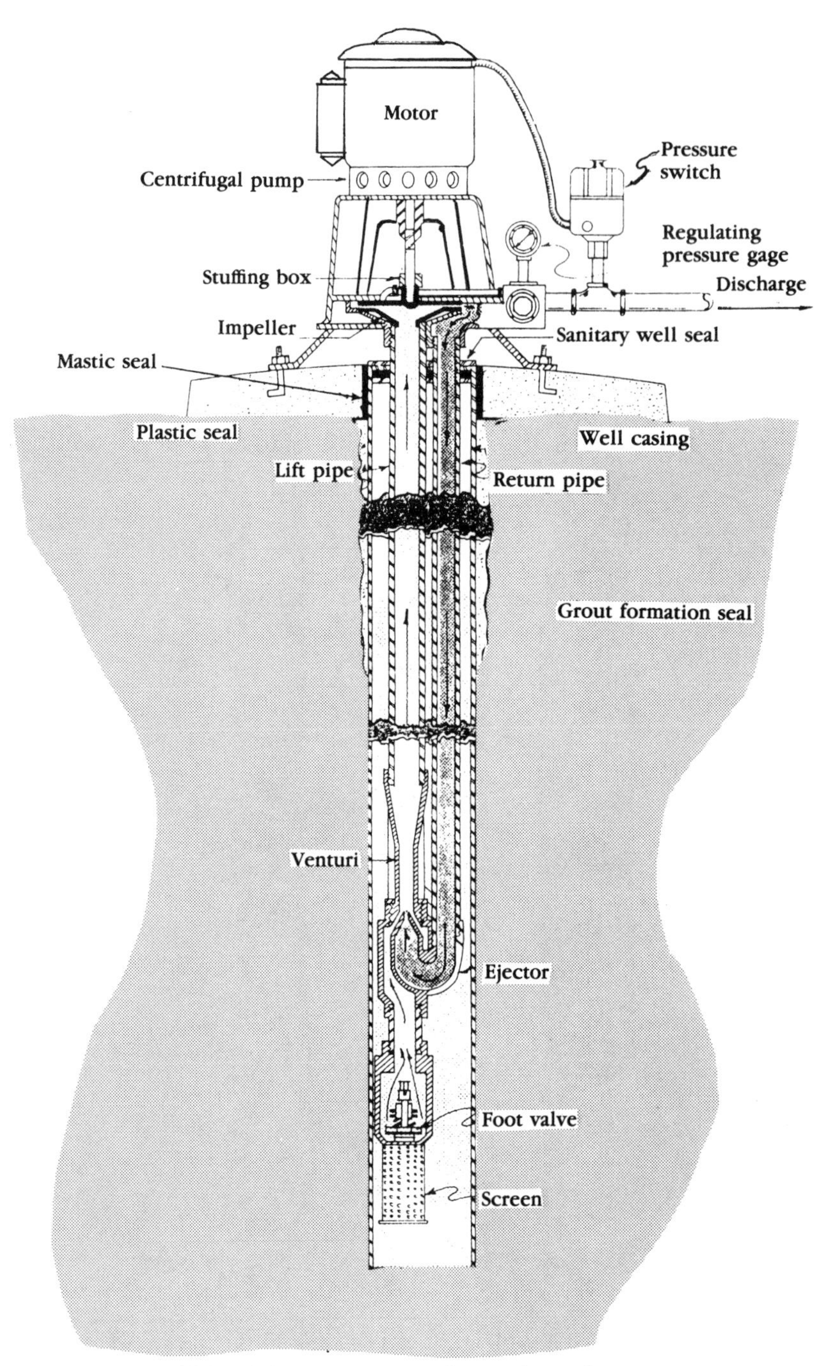

21-3 A jet pump mounted over the well.

from the venturi nozzle pushes it upward where it is captured and lifted by suction. The centrifugal pump then forces the water into the piping.

Pump selection

The following factors should be considered when selecting a pump:

- yield of the well or water source
- daily needs and peak demand of the user
- usable water available in the storage or pressure tank
- size and alignment of the well casing
- total operating head pressure of the pump at normal delivery rates, including lift and all friction losses
- difference in elevation between ground level and water level in the well during pumping
- availability of power
- ease of maintenance and availability of replacement parts
- first cost and economy of operation
- reliability of pumping equipment

If the well yield is sufficient, a pump capable of delivering the peak demand should be installed. If the yield is low by comparison with the peak demand, commensurately increased storage is in order. The interval between starting and stopping the pump should be as long as possible, and never less than 1 minute. Rapid cycling destroys pumps and motors.

Remember that any pump that cannot be wholly submerged during pumping must depend upon suction to raise the water from the well. The suction lift, which is the distance from the surface of the water source to the center axis of the pump, can be no more than 15 to 25 feet, depending upon pump design and altitude of the pumping site above sea level.

In all cases the best bet is to obtain thorough, competent advice from an expert when selecting a well pump. Factory representatives from any of the name-brand pump companies are usually the best qualified to make recommendations.

Pump installation

The details of installing pumps varies widely and depends primarily upon the kind of pump used. The common arrangements are for the pump to be fully submerged, mounted atop the well casing, mounting atop a water storage tank at a remote location, or set separately from either the well or a storage unit, perhaps in a pumphouse or perhaps in a basement.

Whatever the situation, it is imperative that the installation be properly made according to the manufacturer's instructions. That is the only way to ensure proper and trouble-free operation, realize decent service life and keep any warranties in force. Above all, the installation must be a sanitary one, made in such a way that there is no chance for any contamination from ground water or any other source to get into the water supply.

In most instances it is a good idea to let a qualified well service worker make the pump installation, especially if the pump is submersible, or a wellhead vertical turbine.

Pumphouses and pits

There was a time when the pump pit was a popular way to house the well pump and associated equipment. This consisted of making an excavation at the well head to below frost level, lining it with a large-diameter section of concrete pipe or pouring a concrete box, and fitting it with a cover. If the pump itself was located elsewhere the pit was for well access and was buried under 2 feet or more of earth. Otherwise a cover was fitted at or slightly above grade level. Such arrangements involve too much hazard of pollution or contamination to be recommended under any circumstances, and many states will not allow their construction.

Pumphouses (FIG. 21-4) are a different matter. This is an excellent arrangement, especially where the pump will be mounted atop the well casing or can be located next to it. This keeps the machinery and attendant noise out of the house and makes maintenance and servicing a bit easier.

The pumphouse floor area should be amply large enough to provide plenty of working room around the equipment, plus some storage space and a spot for any water treatment equipment that might be required or contemplated. If the location is rural and susceptible to long power outages, consider making an allowance for the installation of an emergency generator of sufficient capacity to power the pump motor. An alternative would be a larger unit, located there or elsewhere, sufficiently large to take care of other household needs as well.

If possible the pumphouse floor should be approximately centered on the well, made of poured concrete that is sloped away from the wall casing in all directions. The walls can be masonry but are more often wood frame. The roof is also typically wood frame in the shed design; depending upon the kind of pump involved, there may have to be a sizable hatchway in the roof directly over the well to allow for withdrawing the pump from the well. In some designs the entire roof is removable and sometimes the walls can be taken down as well. The pumphouse should be well insulated, commensurate with the local climate, and provided with a means to keep the interior at no colder than 40°F. A couple of 60-watt light bulbs or a small, thermostatically controlled electric heater will do the job, or self-regulating heat tape can be wrapped around the pump and piping.

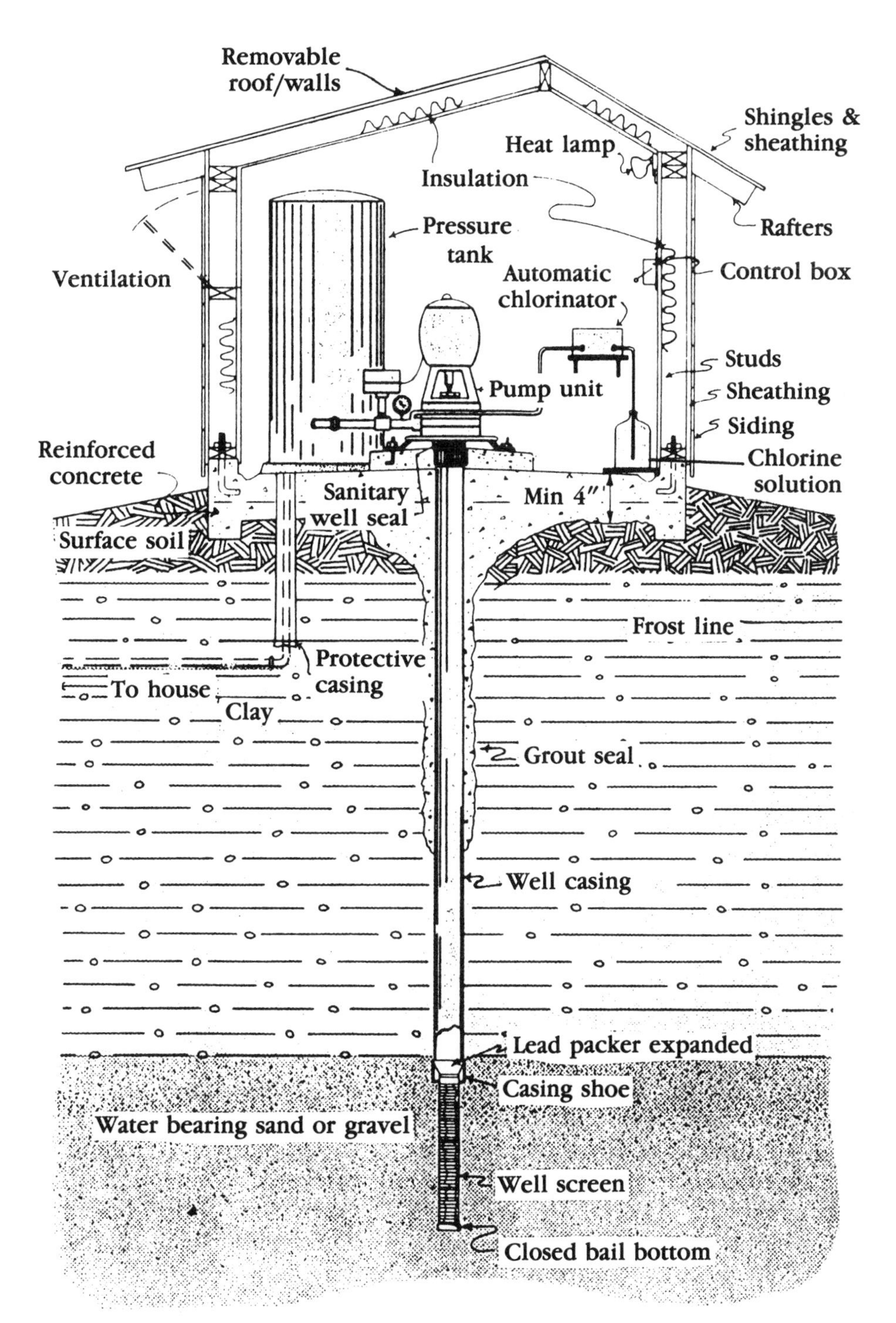

21-4 A typical pumphouse arrangement, featuring removable walls and roof, freeze protection and ample room for equipment.

Pitless adaptors

When the pump is submersible, located in the well itself, or the pump is a type that can be located in the basement of the house, there is no need for a pumphouse. Instead, a *pitless adaptor* (FIG. 21-5) can be installed in

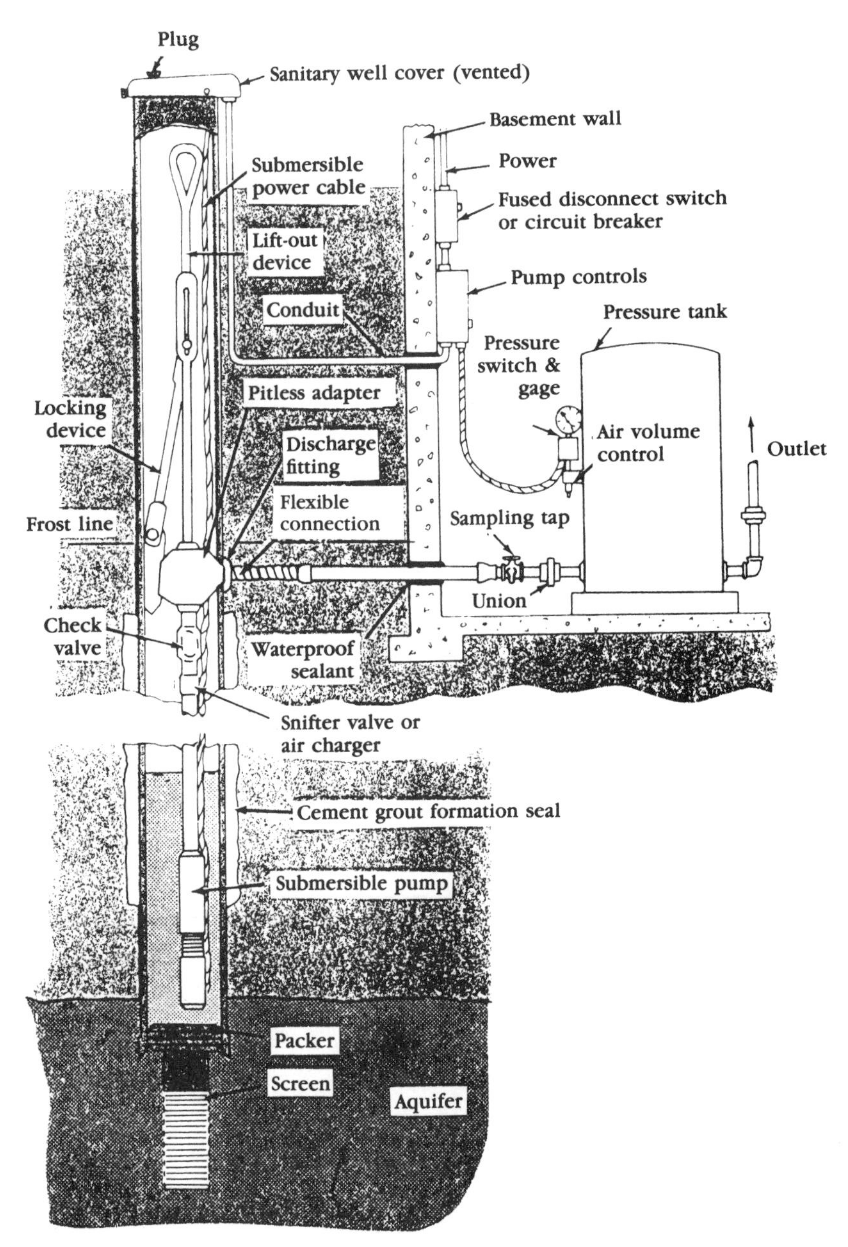

21-5 A pitless adaptor installed in combination with a submersible pump and basement storage. Other kinds of pitless adaptors are available for different types of pumps and installation conditions.

the well casing a suitable distance below frost line. There are several kinds of these devices and many brands, designed so the discharge or suction line can be joined directly to the well pipe. The casing is continued to a foot or so above grade and capped with a *sanitary well cover.* This provides ready access to the pitless adaptor and the well and eliminates any need for a well pit.

Not all models of pitless adaptors are of good quality or design, and not all are acceptable in all locales. Be sure to check what is locally available and acceptable. None are easily installed, and welding, case-cutting with special tools, cutting by torch and power threading are likely to be part of installation. Generally the services of an experienced professional are required. These installations frequently must be inspected and approved by local authorities, and they must also be pressure-tested in certain ways. In some areas the state and local health authorities can supply a list of licensed and certified contractors who are authorized by law to construct wells and install pumping systems.

DISTRIBUTION

The distribution portion of a water system consists of the pipes, fittings, valves and associated gear required to get the water out of the well and into the house. It does not include the domestic supply lines that travel throughout the house. In the typical individual water system, distribution is accomplished through a single underground pipeline from the well, pumphouse or storage cistern into the house as far as the storage tank, pump if located there, or to a connecting point with the domestic supply line.

Pipe and fittings

Many kinds of pipe are suitable for water distribution systems. Those most commonly employed in residential installations are galvanized iron or steel, plastic and copper.

Galvanized pipe tends to corrode rather rapidly, especially where corrosive soil or water prevails. It is particularly susceptible to rotting through at threaded ends, where the threads expose raw metal and also reduce the pipe wall thickness. It also tends to clog and choke up readily with rust and scale under certain conditions. However, it can be covered with a protective coating, such as a bituminous compound, as it is laid, to extend its service life.

Copper pipe or, more often copper tubing, is a popular choice. The initial cost is high but copper has a considerably longer useful service life than galvanized pipe. Corrosion is not a great problem, and the flow characteristics are very good—scale or rust do not build up in it as in steel pipe. Copper is required as a service main in many locales, and is the pipe of choice if it must pass beneath roads or walks or is buried in rough and rocky terrain.

Plastic pipe is now being installed in nearly all situations where copper or galvanized pipe is not required, and certainly is the do-it-yourselfer's choice. It is lightweight and very easy to handle, simple to install, and has excellent hydraulic properties. The pipe is tough, corrosion- and rust-proof, impervious to practically anything except certain solvents, and neither scale nor rust will build up on the inside walls. Two popular kinds are the flexible, black polyethylene (PE) that comes in long-length coils and the white polyvinyl chloride (PVC) that comes in semirigid lengths. Whichever is used should be certified by an acceptable testing laboratory as being nontoxic, nontaste-producing and safe for carrying potable water.

Fittings for galvanized iron or steel pipe are made of the same material. Those for copper may be of copper or brass, depending upon circumstances. Fittings for PE pipe may be nylon or some other rigid, reasonably hard plastic, while those for PVC will be of a like material. Valves are mostly cast from brass or bronze, although some plastic valves are beginning to be installed.

Installation

Fittings should always be made up in the manner approved by the trade for each combination of fittings and pipe. This typically includes pipe dope or non-stick tape and stoutly wrench-tightening for galvanized threaded assembly; approved cleaner, primer, and welding solvent for semirigid plastic; double stainless steel worm-drive clamps for push-in fittings on PE pipe; and soldered or mechanical fittings for copper lines.

Pipes should be laid as straight as possible along the center lines of trench bottoms. Make curves as gentle as possible. It is good practice to lay several inches of sand on the trench bottom and compact it, then bury the line with another several inches of the same. Many locales require this or a similar arrangement. Backfill must be clean and free of rocks or trash. Trench depth must be below frost line to avoid winter freezing. In warm climates, laying the pipeline at least 3 feet below grade will help keep the water in it cool during the hot months.

Pipe capacity

Water pressure at the most remote fixture of a household water system can be as low as 20 pounds per square inch or as high as 60 or 70 psi. The former creates an exasperatingly slow stream of water from the tap while the latter leads to explosive sprays and also can cause problems with garden hoses and automatic valves in washers, among other things. About 40 to 50 psi is generally considered satisfactory. The pipeline should be adequate to deliver good pressure at the required peak flow.

One way to determine the correct pipe size is to consult FIG. 21-6. This information is based upon standard galvanized pipe, but because the hydraulic properties of plastic and copper piping are somewhat better, no loss will be incurred by applying the information to those pipe types as well.

In this chart, pipe size is shown as a function of head loss H in feet per foot, pipeline length L in feet, and peak demand in gallons per

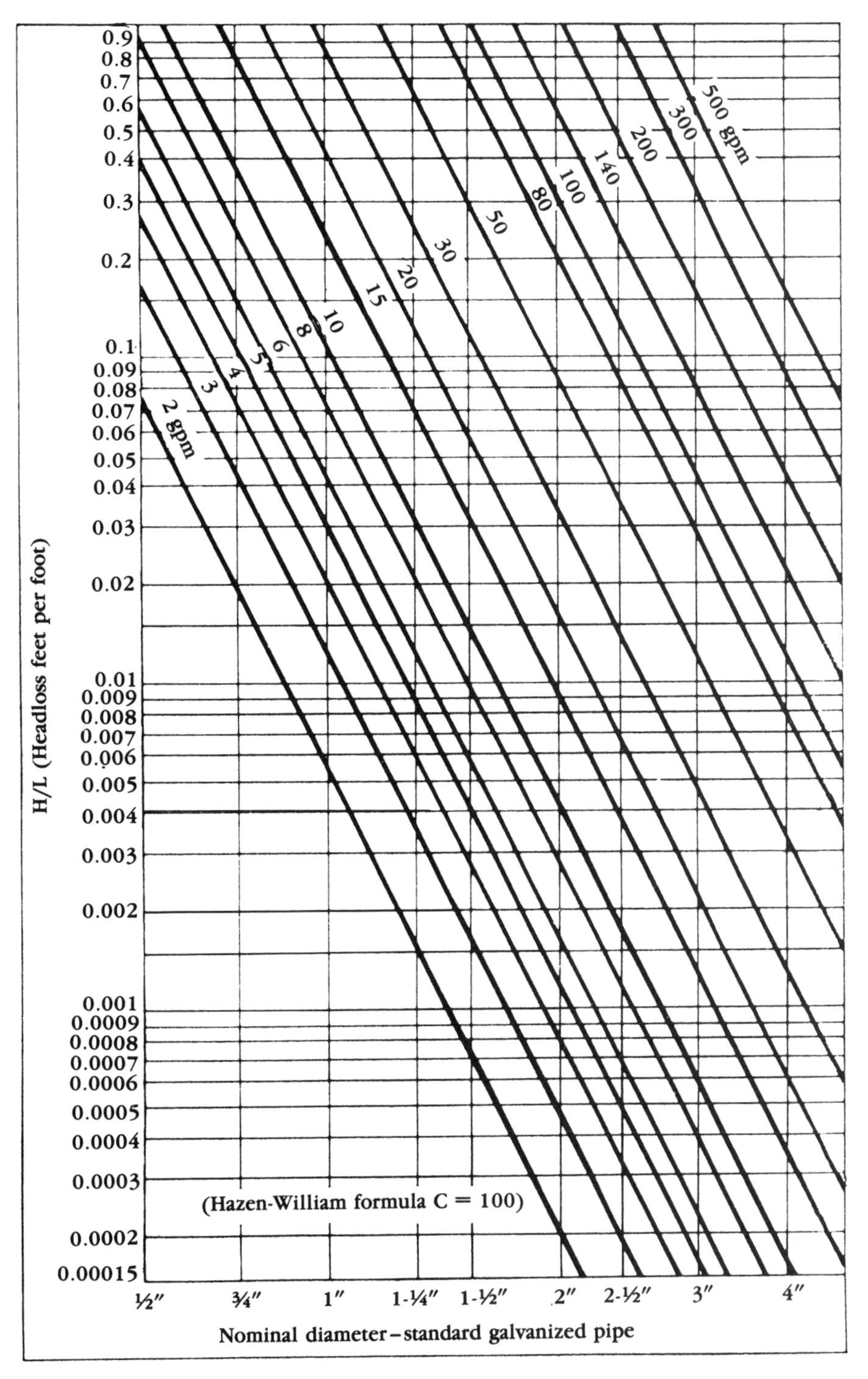

21-6 Head loss versus pipe size.

minute. For example, suppose that the supply pipe runs 150 feet from a well to a house, and the pump must force the water upward 24 feet at a minimum entrance pressure of 40 psi and a rate of 20 gallons per minute.

To find the operating head loss, which is the amount of pressure drop through the length of the pipeline from various causes, the first step is to determine the total head. This is the elevation head or lift plus the required pressure head at the service. The pressure head is calculated on the basis of 1 pound per square inch produced by a column of water 2.31 feet high, and the elevation is given as 24 feet. Thus:

$$H = 24 + 2.31 \times 40 = 116.4 \text{ feet operating head loss}$$

To find the head loss in feet per foot of pipeline, also called the hydraulic gradient, divide the head H in feet by the pipeline length L in feet.

$$H \div L = 116.4 \div 150 = 0.776 \text{ foot per foot}$$

The peak demand is given as 20 gpm, and the gradient is 0.776, or about 0.78. Follow the 20 gpm line of the chart upward to 0.78 H/L and locate the crossing point, then read out the pipe size at the bottom of the chart. In this case a diameter of about ⅞ inch is indicated. That is not a standard, so the first standard trade size larger, 1-inch pipe, would be selected. That, in fact, is a commonly used size for domestic well service.

The water source is not always below the point of delivery into the house. When it is above, gravity aids in supplying the water. Let's assume that a storage cistern is 75 feet above the delivery point and the pipeline is 200 feet long, a minimum pressure of 30 psi is required and the flow rate must be 15 gpm. Again the first step is to find the maximum operating head loss. Now, however, that equals the difference in total head and required pressure head at the service. Thus:

$$H = 75 - 2.3 \times 30 = 6 \text{ feet}$$

To find the head loss in feet per foot:

$$H \div L = 6 \div 200 = 0.03 \text{ foot per foot}$$

Consulting the chart, the head loss of about 0.03 in the H/L column reads out to a pipe size of just over 1½ inches, and that would be the optimum choice.

The pressure given is for the entry or connection point of the distribution pipeline to the internal domestic supply. The pressure obtained at a water tap in the third floor rear bathroom could be much less, depending upon the complexity of the internal piping system. It is necessary to decide upon a minimum acceptable water pressure at the farthest point of use, then determine the head loss through the internal system between that point and the entry point. Then the required entry pressure can be established to provide that minimum pressure.

Additional head losses will be incurred by including fittings in the pipeline. These losses are not especially important in pipelines longer than 300 feet. In short runs, especially shorter than 100 feet or so, they may be consequential. Usual practice is to make an allowance in the calculations for such losses on the basis of added pipe length. Consulting TABLE 21-1, you can see that adding five couplings, two 45-degree elbows, two 90-degree elbows and a gate valve to the system just discussed is the equivalent of adding almost 15 feet of pipe to the line.

21-1. Pipe-length Friction Loss Allowances for Valves and Fittings

Diameter of fitting (inches)	*90° std. ell (feet)*	*45° std. ell (feet)*	*90° side tee (feet)*	*Coupling or straight run (feet)*	*Gate valve (feet)*	*Globe valve (feet)*	*Angle valve (feet)*
3/8	1	0.6	1.5	0.3	0.2	8	4
1/2	2	1.2	3	0.6	0.4	15	8
3/4	2.5	1.5	4	0.8	0.5	20	12
1	3	1.8	5	0.9	0.6	25	15
1-1/4	4	2.4	6	1.2	0.8	35	18
1-1/2	5	3	7	1.5	1.0	45	22
2	7	4	10	2	1.3	55	28
2-1/2	8	5	12	2.5	1.6	65	34
3	10	6	15	3	2	80	40
3-1/2	12	7	18	3.6	2.4	100	50
4	14	8	21	4	2.7	125	55
5	17	10	25	5	3.3	140	70
6	20	12	30	6	4	165	80

STORAGE

There are several approaches to water storage in small individual water supply systems. The most commonly used arrangement is a pressure tank located either in the house basement or in a wellhouse. The tank size may be selected on no particular basis other than that which happens to be readily available, or it may be of a size and shape that will conveniently fit in a designated spot, or because it is inexpensive. However, there are better and more definitive ways to determine an appropriate size. One uses the number of baths in the home, a normal 7-minute peak demand and pump capacity to arrive at a size recommendation. The storage capacity varies with the type of tank—precharged, supercharged or standard. Consult your dealer for design details and recommendations.

The problems with pressure tank storage is that their capacity is usually small when compared to the total daily household water demand. While a good-sized tank might hold 80 or 100 gallons, demand could run to several hundred gallons daily. If fire-fighting capacity is desired, storage against water shortages or power outages is part of the plan, or a low-yield well needs plenty of recharge time, a much larger storage facility will be needed. Size can be calculated on the basis of peak water needs for a particular period such as a week or a month, or on an arbitrary gallon figure like a standard 1000- or 5000-gallon tank size.

Storage reservoirs can be constructed of concrete, concrete block or brick right on site, and usually are partly or wholly buried in the ground. Commercially made steel or fiberglass tanks are available for the purpose and are usually buried. Above-ground tanks may also be made of redwood heartwood or bald cypress. No storage unit should be coated on the inside with asphalt or any other substance that might foul the water, but they can be treated on the exterior. Masonry units are parged with cement plaster, and this can be done both inside and out. Certain paints are available for interior coatings: check with your local health department for current information and recommendations.

Whatever the construction, the tank should be made tight and so as to prevent any possibility of contamination from outside sources. Figure 21-7 shows a typical concrete reservoir with screened inlet and outlet. The sanitary manhole cover should overlap by 2 inches a curb that stands at least 4 inches above the reservoir top. Before you build, check with local authorities to find out if there are any construction requirements in force or permits needed in your locate for private water storage facilities.

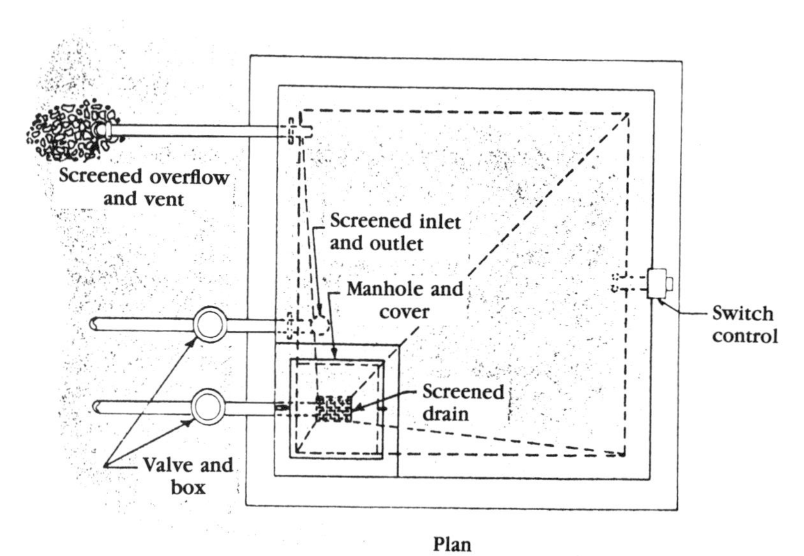

21-7 Typical poured concrete water storage facility.

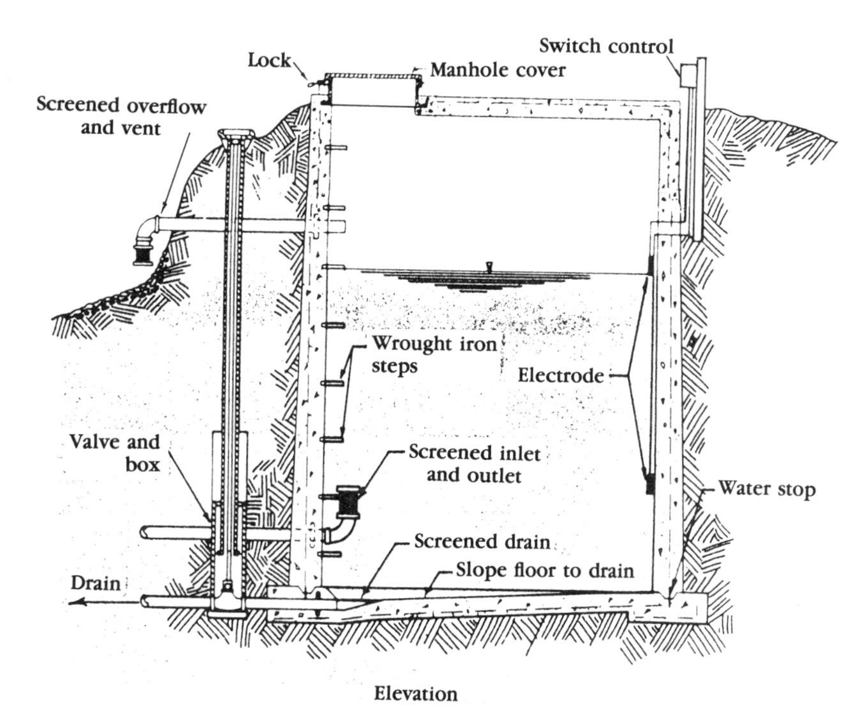

21-7 *continued*

Glossary

Aeration, zone of Formation extending from the water table to the surface of the earth, in which no free water is found.

Aerobic bacteria Bacteria that cannot thrive without oxygen.

Anaerobic bacteria Bacteria that thrive in the absence of oxygen.

Annular velocity Speed of air or water through an annulus.

Annulus Circular space between a central pipe and a surrounding casing.

Bailer A special tool used to remove sand and cuttings from a bore hole.

Bentonite A kind of clay used in drilling.

Black water Water from a toilet.

Bottom-hole drill A drill that is operated by a motor that fits into the bore hole.

Cable-drill tool A drilling tool that is raised and lowered in the bore hole on the end of a wire cable.

Casing shoe An edged collar fastened to the lower end of a well casing that is to be driven into the earth.

Centrifugal pump A pump that drives water by means of the rotary action of its blades.

Cuttings Chips of the formation produced by the drill bit.

Dart valve A fast-acting valve sometimes installed on the bottom of a bailer.

Development The process of removing fine sand and silt from the area immediately surrounding the well screen.

Down-hole hammer A rotating, reciprocating bit driven by an air hammer (both are lowered to the bottom of the bore hole).

Draw down The lowering of the static water level by pumping the well.

Drill stem One of the components of the drill string.
Drill string A series of tool parts of varying kinds connected together end to end to accomplish the drilling.
Drive pipe Pipe attached to a well point.

Effluent The liquid that emerges from a septic tank.

Formation A general, geological description of the subsurface earth and stone.

Gray water Sewage from sinks and tubs that does not include water from toilets.
Grout A cement slurry, sometimes containing sand as well.

Head drive An arrangement wherein the motor driving the drill is mounted atop the mast.
Hydraulic rig When the drive motor is actuated by hydraulic power.

Impeller pump A type of pump that moves liquid by means of one or more internal rotors.

Jar A tool section of two loose-fitting parts, incorporated in a drill string to produce an upward hammering action as the cable is dropped and then immediately raised.
Jet lance A type of well-drilling tool through which water is forced.
Jetting The process of using a powerful stream of water to bore a hole into the earth.

Kelly A shaped or splined steel bar that passes through and is turned by a rotary table on the drilling rig.

Lead packer A shaped ring of lead that is fastened to the top end of a well screen, used to seal the screen to the casing.

Mast A vertical tower on a drilling rig that supports the tools.
Meander tank A septic tank designed to provide the longest practical path for the effluent.
Mud scow A tool used to excavate a bore hole in a soft formation.

Packing, artificial The gravel placed around a well screen.
Packing, natural The gravel around a well screen after the well has been developed.
Pathogen Harmful organisms invisible to the naked eye.
Pitless adapter A device for connecting a water line to a well that requires no pit and allows above-grade access.
Percussion drilling A system of raising and lowering a drill string on a cable so that the bit pounds a hole into the formation.
Pull down A method of applying downward pressure to a casing to force it into the formation.

Recharge Rain or other water that seeps into an aquifer.
Rig Equipment used to drill or bore wells.
Riser Special pipe connected to a well point.

Rotary table A flat, circular steel table with a shaped hole centered in it to accept the kelly, for driving a drill bit.

Sand line A light cable used to raise a bailer.

Sand pump A pump mounted inside a bailer to lift sand from the bottom of a bore hole.

Saturation, zone of The area in an aquifer that contains free water.

Seep A natural spring just a few inches below the surface of the earth that keeps an area constantly wet.

Siphon tank A septic tank containing a siphon for fast movement of the effluent from one tank section to another or to the drain pipe.

Spudding The up and down action of a percussion drill; the mechanism providing that action.

Static water level The water table when the well is not being pumped.

Suction head Total negative pressure required to lift water from the source to the pump center, expressed in feet.

Termie A pipe passed down an annulus, and through which gravel or grout can be forced to the bottom of a bore hole.

Vacuum pump A pump that raises water by suction (vacuum).

Wash down To use a stream of water under pressure to provide a hole in the earth under a casing, so lowering it into the earth.

Water table The upper surface of groundwater within an aquifer before any pumping is done.

Well casing A steel pipe that might contain a second pipe within itself, to carry water; the casing alone may also carry water.

Well pipe The pipe that contains and carries the well water to the supply line.

Index

J

L

M

N

O

P

R

S

Z